T0271022

Safety in Industry

Learn from Experience

Safety in Industry
Learn from experience

Brij Mohan Bansal

CRC Press
Taylor & Francis Group
Boca Raton London New York

CRC Press is an imprint of the
Taylor & Francis Group, an **informa** business

WOODHEAD PUBLISHING INDIA PVT LTD

New Delhi

First published 2024
by CRC Press
4 Park Square, Milton Park, Abingdon, Oxon, OX14 4RN

and by CRC Press
2385 NW Executive Center Drive, Suite 320, Boca Raton FL 33431

CRC Press is an imprint of Informa UK Limited

© 2024 Woodhead Publishing India Pvt. Ltd., 2022

The right of Brij Mohan Bansal to be identified as author of this work has been asserted in accordance with sections 77 and 78 of the Copyright, Designs and Patents Act 1988.

Print edition not for sale in South Asia (India, Sri Lanka, Nepal, Bangladesh, Pakistan or Bhutan).

British Library Cataloguing-in-Publication Data
A catalogue record for this book is available from the British Library

ISBN13: 9781032630120 (hbk)
ISBN13: 9781032630137 (pbk)
ISBN13: 9781032630144 (ebk)

DOI: 10.4324/9781032630144

Typeset in Times New Roman
by Bhumi Graphics, New Delhi

Contents

Preface

There are many books available in the market on safety covering different aspects of safety. In the book, I have attempted to share my experience on safety through some case studies which are explained in a very simple manner, with simple sketches at some places and with basic purpose that even work-force in field can understand the reasons for the accident / Fire and remember them for long time. I have gone through many case studies, where author writes about the chemical process in complicated language, explaining the process, flow-Diagram of the process and what could have been the probable reasons for the accident. While I appreciate their approach but to my mind, the person in field wants to understand the cause and remedies in a very simple language with reasons and learnings. I hope that my assumption will be helpful to all the readers.

I have covered in my discussion and may be seen in these case studies that the reasons for accidents are almost common i.e.;

    i. Design mistake

    ii. Human error / Negligence;

    iii. Lack of training / knowledge;

    iv. Emphasis on production nos;

    v. Poor maintenance practices;

    vi. No standard operating procedures (SOP);

    vii. Non-availability of proper PPEs

    viii. Instrument failure;

    ix. Over confidence;

    x. Person not medically fit.

Another salient observation is that if the mistake is corrected or fire is tackled at the initial stage, the major accidents can be avoided. To achieve that, people in field are required to have knowledge about do's and don'ts, knowledge on Standard Operating Procedures (SOP), Emergency handling and use of PPEs. Sometime common sense is essentially needed to understand the problem and act for its immediate control.

My key word for safety is "Training and Retraining". At the time of emergency, if the person(s) on spot is alert and knows the basics of safety, he can handle the situation very confidently and minimise the loss of property and human life.

I have started with introduction to basics of safety, operational safety and in short, given 32 case studies which I am sure will be of great interest to all the readers. At the end of the book, I have given some tips to be remembered by all as thumb rules for being safe and for safety of our colleagues and the property. These tips are based on my knowledge and experience and they may need to be modified as per the site condition.

I expect this book to be useful for the people working in Industries. Your feed-back will be of great value to me.

I am thankful to all my friends and well wishers who gave me motivation to write this book. My special thanks to Mr. Mr. Shashi Vardhan Pandey and Mr. Sunil Kumar, ex-General Managers (HSE), Indian Oil Corpopration Ltd., for their active support & guidance in writing this book.

The contents in the book are based on my experience in oil industry, various web sites, OISD guidelines, OSHA standards and published case studies, etc. I pay my respect and thanks to all of them for helping in this compilation of information in a book form. In the bibliography given at the end of the book, I have humbly acknowledged all the references and any missed inadvertently, would like to seek pardon for the same.

Regards

B. M. Bansal
Email: bansalbm51@gmail.com

Reader's views

I have gone through the book *Safety in industry*. Learn from Experience written by Mr. Brij Mohan Bansal, former Chairman of Indian Oil Corporation Limited. Mr. Bansal has had a long and diversified experience in the Oil Industry. In this book, he has explained very will the importance of near-miss incidents and covered a good number of case studies in as simple a manner as possible. He has elucidated the lessons learnt and given his valuable recommendations thereof. Based on a world-wide-experience, he has even, clearly and succinctly, listed the vital tips for maintaining safe working conditions in the industry.

I think this book will prove very useful for the workforce as well as the management of the industrial sector. Safety in industry is likely to even demonstrate a practical aspect of working in industry to students of engineering who will be entering the industry in the near future.

Prabh Das
Managing Director &
Chief Executive Officer

Reader's views

Truly the first safety book of its kind.......

Industrial safety refers to the safety management practices which apply to the industrial sector for safe-guarding workers, machines, plants and equipment ,buildings, structures and the environment.

Being myself,a chemical engineer experienced in handling all sorts of chemicals,I have read several books on safety in the past but this one written by B. M. Bansal is truly the first of its kind.

Bansal gives to the readers his entire life span of 40 years of experience,from a trainee engineer rising to the CMD of IOC, a Fortune 100 conglomerate, in a very simple but lucid and effective manner through:

- **17 major reasons** of most accidents, expertly culled out using his sieves of experience
- **32 case studies** with learnings from each of these pearls of wisdom. Very costly for those who underwent these cases but presented to the readers in a very easy- to- grasp manner
- **55 golden tips** to maintain a good and healthy safety system

I strongly recommend this book titled **"Safety in Industry. Learn from Experience"** for everyone who is part of the industrial environment to study and imbibe the valuable lessons imparted by Bansal.

Sincerely yours,

Vijay Kumar Soni

vksoni@gmail.com

Director GFCL EV Products Ltd;

Director GFCL Solar & Green Hydrogen Products Ltd;

Head of Projects & Key Initiatives, GFL;

Director GFLGM Fluorspar Co., Morocco;

Director Swarnim Gujarat Fluorspar Ltd ;

Director SCC Consulting India Pvt Ltd

First job in paper mill: Tea every two hours

After having done my B.Tech in Chemical Engineering from IIT, Delhi, my first job was in a paper mill. The mill was located near a bamboo forest in a remote area of Orissa. Bamboo is the essential raw material for the manufacturing of paper. Shredded bamboos are cooked with caustic in the 'Digester' machine, which gives off quite an unpleasant smell, and it took me several days to get used to it.

In the 'Soda Recovery Unit', the 'Black Liquor', emerging from the combined bamboo and caustic soda, is then fired in a boiler, where liquid is burnt and sodium recovered in molten form. The air in the plant would be very dusty and full of soot, especially during the soot blowing operation of the economizer coils. It would be so bad we would be forced to run away to a faraway spot and gulp down cups of hot tea to soothe our choking throats.

I was aghast: there was no formal safety department in the mill except for the firefighting facility.

Decision to leave paper mill

Since the paper mill was in a remote place, I had no intention to continue there for long, anyway.

When I received the offer from a major oil company, I left for Delhi by an early morning train the very next day. It was a circuitous journey, first coming to Raipur, then to Nagpur, and finally Delhi, taking almost two days to complete the journey. Since it was the first time the paper mill management had recruited engineers from IIT as Management Trainees, they threatened to take action against me for jumping the bond. I did not budge however on my decision, and with time, the matter cooled down.

First impression of petroleum refinery

In college, our impression of petroleum refineries was that these would be dirty places with puddles of oil everywhere. When I entered the refinery however, I was pleasantly surprised to see a very neat and tidy plant area. In college, our knowledge was limited to simple distillation columns, with single reflux and top and bottom products only. Here I learnt that to get various products from the same column, there are number of side-draws and circulating refluxes to the columns for better fractionation. In chemical engineering, a general concept about design of columns, exchangers and other equipment is taught. A special course on petroleum technology was arranged by the oil company for the entire batch of chemical engineer trainees. Being in a hazardous and disaster-prone industry dealing with Oil and Gas, they knew how essential it was to lay special focus on safety aspects.

Exposure to safety norms

The oil company had an excellent schedule for training the new recruits. For the first few weeks, we were given an orientation to the company, its policies, plant operations and safety and fire-fighting training. We were explained the dos and don'ts of the plant area and the uses of various safety equipment, including the types of fire extinguishers and where each one was to be used. In practical training, we had to learn to extinguish the fire in a small oil tank, kept specially for demonstrative purposes. I was told to extinguish this fire by throwing foam on the flames. I was to aim at the oil layer so that the foam floated on top of it, cutting off air supply, and hence extinguishing the fire. Trainees were exposed to unit operations for a few months to get a hands-on experience. Safety shoes, helmets and goggles were issued for use inside the plant. The Fire and Safety Department was headed by a dynamic Sikh gentleman who was fondly nicknamed 'Safety Surd'. Later, he procured a good job somewhere in the Middle East. The department circulated safety instructions, kept a record of the hot jobs going on in the battery limit, inspected reportedly unsafe conditions in the plant area and carried out investigations in case of any accident or fire. Fire-fighting was required only once in a while, but they were supposed to be always ready to reach the danger spot in the least time possible.

Safety – whose responsibilty

Safety plays a very important role in maintaining the reputation, sustainability and growth of an organization. Frequent accidents / fires / fatalities in any industry lead to de-motivation amongst the employees, tarnish the image in the community as well as in the eyes of other stakeholders, apart from paying huge compensation and penalties (while the direct losses may be only the tip of the iceberg).

Safety is everyone's responsibility, but the drive has to come from the top. The top management has to formulate the safety policy, prepare safety manuals and distribute to all supervisors and control rooms, provide the requisite number of Personal Protective Equipment (PPE) and train and retrain on the proper use of PPEs through the safety department, bring awareness among employees through safety talks and departmental safety meetings. Management has to introduce work permit systems, incident reporting, analysis and investigation procedure, emergency handling plans and mock drills, safety inspection and safety audit, Hazop studies and monitor the action plan to liquidate the recommendations of such audits.

Even if all the above steps are taken, unless safe habits are inculcated among the employees as a safety culture, unsafe operations will still be prevalent. Hence, it is very important to develop a safety culture in the organization by everyone following the safety guidelines at all times. The top and middle management play a very crucial role by setting examples and keeping a close watch on near-miss accidents.

Safety: Not a one-time assignment

Safety is not a one-time assignment, and people have to be always alert. In fact, safety starts from home. Once it is a culture, one can locate unsafe activity wherever it is and take immediate corrective action. Safety cannot wait. A company might have prepared nice manuals and instruction books or obtained a high safety rating from international agencies. However, safety performance cannot be improved unless there is awareness in the field and a push from management to ensure implementation and follow-on guidelines. A small fire, if controlled at the initial stage, can avert a major fire, but if the operator chose a wrong extinguisher to fight this small fire, or if he does not know its operation or suppose the extinguisher is not charged after its use in past, the small fire will turn into a major one.

Hence, Training and Retraining of the manpower on a regular basis is a must for any industry, if it is to be believed safe. Mock drills help in understanding role-play during an emergency and to ensure the readiness of all the equipment and alertness of the employees. Frequent safety audits (even by third parties) bring out the weaknesses in the system, and their redressal through a strictly monitored action plan is the responsibility of the management.

Even at the design stage, we should follow the guidelines laid down in manuals, standards and suggested by bodies like Oil Industry Safety Directorate (OISD). Hazop and Hazard analysis must be carried out to take care of design mistakes in the implementation stage of the project. Before commissioning, a proper safety audit is a must from the operational safety angle.

Construction – Safety is another important part of safety culture. Training of contract labour, supply of PPEs and strict implementation of these as well proper precautions are to be taken for working at heights, as well as while working on pipelines and excavation work.

Near-miss incidents – not to be ignored

Minor incidents like slipping of a person, stumbling against some obstruction in the path, losing balance, etc., keep happening in industry, but these are termed as near- miss incidents, which could have become serious accidents. Such near- misses should not be ignored as worldwide analyses show that if these incidents are not analysed properly and corrective action is not taken immediately, these can result in major accidents and cause casualties.

More than forty years' industry experience

In my long working experience of more than forty years, I have worked as a production engineer, section head, Departmental Head, Refinery Head and finally as Director / Chairman / CEO. More than twenty years were spent in refineries and among other functions, safety was one of the key focus areas. My philosophy on safety was that "we must learn from others' mistakes, as to learn from our own is very expensive". Hence, apart from the safety instructions and advice, company should circulate the findings and learnings of accidents happening in India and the world over. In training, videos of accidents are very useful, as what we see remains fresh in the mind for a long time. Senior executives should take rounds in the field and develop rapport with employees to understand their knowledge on safety. They have to understand which safety issues are still pending and how to improve the safety of both employees and plants further. These executives have to demonstrate their seriousness about safety by wearing PPEs while moving in the field and ensure that everyone in the field is using PPE as per the site requirements.

From my experience in the industry, I have come across and reviewed the cases of many accidents and carried out safety audits of a number of refineries and projects. One of my observations was that the shortcomings in these accidents were similar in nature. The main reasons for the accidents have been

- Design mistakes
- Human Errors / Negligence
- Lack of knowledge
- Lack of management's focus on safety (Priority to production maximisation and profit, neglecting proper inspection and maintenance of the equipment)
- Non-availability of standard PPEs and training to staff for proper use
- Non-availability of Standard Operating Procedures (SOPs) and Emergency handling guidelines
- Carrying out some activity without proper permit or non-compliance of safety conditions stipulated in the work-permit
- Over confidence or taking action without understanding the risk
- Person not medically fit to work at height / in confined space
- Instrument failure or leakage due to defective metallurgy or gasket
- Negligence in ensuring proper line-up after maintenance work
- No proper instructions to the field operators or briefing by the shift reliever
- Not giving importance to abnormal behaviour of the equipment or vibrations in pipelines, etc. (predictive and preventive maintenance is neglected).

In the following pages, I would like to narrate some incidents and my learnings from them. These incidents are from my own experience and reviews during my forty years. The analysis and recommendations are as I see them, and they are only to give broad guidelines. They may have to be revised to some extent on case-to-case basis, depending on the actual situation and type of industry.

Chapter 1
Basics of safety and safety management system

The purpose of this chapter is to understand the basics of Safety and Safety Management System and the process of risk reduction in the workplace. The importance of reporting and analysing near-miss/sub-standard acts/sub-standard conditions in an accident-prevention programme has also been discussed.

Chapter 1
Basics of safety and safety management system

1.1   Safety

Safety may be defined as 'control of accidental loss' or 'freedom from accident'. The definition relates to injury, illness, and damage to anything in the occupational and external environment. Here, the term 'loss' means harm to people, damage to property, equipment and/or environment.

1.2   Incident

An event which could or does result in unintended harm or loss may be defined as 'incident'. This includes accidents as well as near-misses. All accidents are incidents, but all incidents are not accidents.

1.3   Accident

It may be defined as 'an event which results in unintended harm or loss.' This includes anything in the work or external environment. An accident occurs normally due to contact with a source of energy (kinetic, chemical, thermal, acoustical, mechanical, electrical, radiation, etc.) or substance above the threshold limit of body or structure. The human body has certain tolerance levels or an injury threshold for each form or energy or substance. Normally, the harmful effects come from single contact, such as cut, fracture, sprain, amputation, chemical burn, etc. The harmful effects of repeated contacts are often repetitive motion injuries, cancer, liver damage, hearing loss, etc and are termed illnesses. Illness may of course also be sometimes the consequence of single contact.

    In terms of people, contact may result in a cut, burn, abrasion, dislocation, etc. or interference with a normal body function (asbestosis, cancer, etc.). In terms of damage, it could be property/equipment damage due to fire, explosion, breakage, distortion, etc. or damage to environment in the form of poisonous air or heavy pollution.

1.4 Near-miss incident

A near-miss incident is an unplanned event that has the potential to cause, but does not actually result, in human injury, environment or equipment damage or a severe interruption in normal operation. Only a fortunate break in a chain of events can prevent an injury, fatality or damage; in other words, a miss is nonetheless a very lucky escape, sometimes by a whisker.

1.4.1 Incident ratio study

Various organisations and agencies have carried out comprehensive study and analysis on incident data, and an example of incident triangle is given below. The figures in each block of the triangle may differ depending on the type of sectors, and any available information may depend on country and work culture, but in each survey/study, the basic concept of Incident triangle was found true.

Serious or major injuries are rare events, and many opportunities are afforded by the more frequent, less serious events to take the necessary actions to prevent major losses from occurring. These actions are most effective when considered as near-misses and sub-standard acts and conditions. Many organisations are finding success by focusing specifically on the behaviour which plays a major role in accident causation, particularly those behaviours which have a potential to cause major losses.

A study says (as seen from the figure 1.1) that for every fatal accident, one may get about 30,000 opportunities to address and rectify it. Out of these, no one knows which one will be a serious one. Addressing the bottom of the triangle means that a focus on unsafe acts/practices and unsafe conditions will help in reducing the number of near-miss and first-aid cases. Hence, the possibility of a major accident will be remote.

The lesson to learn here is that the organisation should focus on the base of the triangle to minimise the possibility of major accidents.

Further studies have shown that the major contributor to the base of the Incident triangle is unsafe acts/practices. These unsafe acts and practices create unsafe and hazardous conditions, which lead to accidents. These unsafe acts and practices are directly influenced by human behaviour, so these are called 'at-risk behaviour'. Promoting the right approach and encouraging desired behaviours of the employees and discouraging undesired behaviours in the workplace can be a game changer in accident prevention. This concept forms the basis of behaviour-based safety.

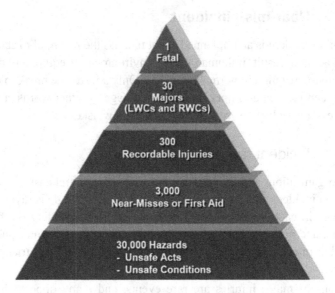

Figure 1.1 Incident triangle

(LWC: Loss Workdays Cases) (RWC: Restricted Workdays Cases)

At-risk behaviours and sub-standard conditions are always symptoms of deeper issues. Examples are given below:

Sub-standard conditions	At-risk behaviours
• **Defects of tools or equipment** ✓ Improper design ✓ Sharp, slippery, worn, cracked, Broken, etc. • **Dress or apparel hazard** ✓ Lack of suitable protective equipment ✓ Improper or inadequate clothing	• **Servicing equipment in operation** ✓ Cleaning, oiling, adjusting, repairing equipment while running ✓ Welding or repairing tanks or containers without purging ✓ Working on electrically energised equipment
• **Environmental hazard** ✓ Excessive noise ✓ Inadequate traffic control ✓ Inadequate ventilation ✓ Improper illumination ✓ Air/water/land contamination ✓ Temperature extremes ✓ Radiation exposures ✓ Poor housekeeping, disorderly workplace	• **Failure to warn or secure** ✓ Failure to place warning signs, tags, or signals ✓ Releasing or moving loads without giving adequate warning ✓ Starting or stopping vehicles without giving adequate warning • **Making safety device inoperative** ✓ Disconnecting or removing safety devices ✓ Adjusting safety device inadequately

Contd...

Contd...

Sub-standard conditions	At-risk behaviours
• **Placement hazard** ✓ Improperly placed/positioned ✓ Inadequately secured against undesired motion • **Hazards from Inappropriate Guard** ✓ Unguarded/inadequate ✓ Mechanical or physical hazard ✓ Lack of shoring or support ✓ Ungrounded electric current ✓ Uninsulated electric current ✓ Unshielded/Inadequately shielded radiation ✓ Unlabelled or inadequately labelled materials • **Hazards outside the organisation's work environment** ✓ Defective equipment/materials/ premises of others ✓ Other hazards associated with activities of others ✓ Natural hazards: weather, terrain, animal, etc. • **Public hazards** ✓ Public transport hazards ✓ Traffic hazards	• **Improper use of hands or body parts** ✓ Grasping objects improperly ✓ Using hands instead of tools • **Operating or working at improper speed** ✓ Feeding or supplying materials too rapidly ✓ Running/Jumping from elevations ✓ Operating vehicle at an unsafe speed ✓ Throwing material instead of passing it or carrying it • **Improper positioning or posture for task** ✓ Entering enclosed space without clearance ✓ Riding in unsafe position ✓ Moving under suspended load ✓ Exposure to swinging loads • **Improper placing, mixing, or combing** ✓ Injecting, mixing or combining substances/ equipment ✓ Improper positioning of vehicles or materials handling equipment for loading/ unloading ✓ Improper placement of materials which create hazards, such as tripping or bumping • **Improper use of equipment** ✓ Using tagged or obviously defective equipment ✓ Using equipment or materials in a manner for which it was not intended ✓ Overloading equipment or structures • **Other sub-standard practices** ✓ No attention to footing or surroundings ✓ Failure to wear safe personal attire, failure/improper use of available PPE ✓ Horseplay • **Hazardous methods or procedures** ✓ Use of inherently hazardous materials/ equipment/methods/procedures ✓ Use of inadequate/improper equipment ✓ Improper assignment of personnel

1.4.2   Basic causes

Basic Causes are the actual underlying or root causes behind the visible symptoms. These are the reasons why the sub-standard acts and conditions occurred at all. When identified, they permit meaningful incident control. Often, these are referred to as Root causes, Real causes, Indirect causes, Underlying or Contributing causes. This is because the immediate causes (the symptoms, the at-risk behaviours and substandard conditions) are usually quite apparent, but it takes probing to understand or diagnose the basic causes and to get control over them. The basic causes of incidents can be broadly divided into Personal Factors and Job Factors. The example of the same is given below:

Personal factors	Job factors
• **Physical/Physiological capability** ✓ Inappropriate height, weight, size, strength, reach, etc. ✓ Restricted range of body movement/ limited ability to sustain body positions ✓ Sensitivities to sensory extremes (temperature, sound, etc.) ✓ Vision/hearing deficiency ✓ Respiratory incapacity ✓ Other sensory deficiencies (touch, taste, small, balance) ✓ Other permanent physical disabilities ✓ Temporary disabilities ✓ Height sickness • **Mental/Psychological capability** ✓ Fear and phobias ✓ Low learning aptitude ✓ Emotional disturbance ✓ Memory failure/mental illness ✓ Poor judgement/Poor coordination ✓ Slow reaction time • **Physical/Physiological stress** ✓ Injury of illness ✓ Oxygen deficiency ✓ Fatigue due to task load or duration/ Fatigue due to lack of rest ✓ Atmospheric pressure variation ✓ Exposure to health hazard ✓ Exposure to temperature extremes ✓ Drugs/under influence of alcohol	• **Leadership and/or Supervision** ✓ Unclear or conflicting reporting relationship/assignment of responsibility or insufficient delegation ✓ Giving objectives, goals or standards that conflict ✓ Instructions, orientation and/or training ✓ Identification and evaluation of loss exposure ✓ Lack of supervisory/management job knowledge ✓ Lack of matching of individual qualifications and job/task requirement ✓ Performance measurement and evaluation or ✓ Incorrect performance feedback • **Engineering** ✓ Assessment of loss exposure ✓ Consideration of human factor/ ergonomics ✓ Standard, specifications, design criteria ✓ Monitoring of construction ✓ Assessment of operational readiness ✓ Monitoring of initial operation ✓ Evaluation of changes • **Purchasing** ✓ Specifications on requisitions ✓ Research of materials/equipment ✓ Specification to vendors ✓ Mode or route of shipment ✓ Receiving inspection and acceptance

Contd...

Contd...

Personal factors	Job factors
• **Mental/Psychological stress** ✓ Emotional overload ✓ Fatigue due to mental task load or speed ✓ Extreme judgement/decision demands ✓ Routine monotony, demand for uneventful vigilance ✓ Extreme concentration/perception demands ✓ 'Meaningless' or 'degrading' activities ✓ Confusing directions/Conflicting demands ✓ Preoccupation with problems ✓ Frustration/mental illness • **Lack of knowledge and skill** ✓ Experience ✓ Initial orientation training/Updated training ✓ Misunderstood directions ✓ Lack of coaching • **Lack of motivation** ✓ Performance is rewarding ✓ Proper performance is punishing ✓ Lack of incentives ✓ Excessive frustration ✓ Inappropriate aggression ✓ Attempt to avoid discomfort ✓ Attempt to gain attention ✓ Peer pressure ✓ Supervisory example ✓ Performance feedback ✓ Reinforcement of improper behaviour ✓ Production incentive	✓ Communication of safety and health data ✓ Handling of materials ✓ Transporting of materials ✓ Identification of hazardous items ✓ Salvage and/or waste disposal • **Work standard** ✓ Adjustment/repair/Maintenance ✓ Communication of standards ✓ Maintenance of standard ✓ Missing or unclear • **Maintenance** ✓ Preventive /Predictive ✓ Reparative ✓ Assessment of needs ✓ Communication of needs ✓ Lubrication and servicing ✓ Scheduling of work ✓ Adjustment/assembly ✓ Examination of units ✓ Cleaning and resurfacing ✓ Part substitute • **Tools and equipment** ✓ Assessment of need and risk ✓ Human factors/ergonomics considerations ✓ Standards or specifications ✓ Availability ✓ Adjustment/repair maintenance ✓ Salvage and reclamation ✓ Removal and replacement of unsuitable items • **Wear and tear** ✓ Planning of use ✓ Extension of service life ✓ Inspection and/or monitoring ✓ Loading or rate of use ✓ Maintenance ✓ Use by unqualified or untrained people ✓ Use for wrong purpose • **Abuse or misuse** ✓ Condition of supervision ✓ Intentional ✓ Not condoned by supervisor ✓ Unintentional

1.4.3   Lack of control

A few of an organisation's problems can be controlled by individual employees or a group of employees, but the majority of loss-producing events are controllable only through the management system.

1.4.4   Safety management system

A system of managing safety may be inadequate because of too few or ineffective system activities. While the necessary activities may vary with each organisation, the list of elements which covers the Safety Management System may include the following:

1. Leadership and administration
2. Leadership and skill training
3. Planned inspection and audit
4. Task analysis procedure and task observation
5. Behaviour-based safety
6. Emergency preparation
7. Personal Protective Equipment (PPE)
8. Safety rules, work permit system and job safety analysis
9. Accident/incident analysis
10. System evaluation
11. Engineering and management of change
12. Communication
13. Health and hygiene
14. HSE promotion
15. Hiring and placement
16. Material and service management
17. Off-the-job safety

1.4.5   Process of managing risks

Hazard identification: The process of managing risks starts from hazard identification. Various techniques, such as HAZOP, HAZIN, safety audits, incident records of premises and elsewhere at similar plants, brainstorming with field personnel, are used to identify the hazards in the workplace. This helps in preparing a list of hazards existing in the premises.

Risk analysis: Each hazard in the list of hazards is assessed with respect to probability and consequence to assess its risk level. The consequence is given a number (1–5), and the probability is also given a number (1–5). The High Risk, Medium Risk and Low Risk levels can be decided as per the matrix drawn based on consequence and probability.

Risk value judgement: As we know, the risk is multiplication of Consequence and Probability. Each hazard, when estimated from the angle of consequence and probability, gives a value. The higher the value, the more severe the risk. High risks need to be addressed on priority.

Tolerate: On evaluation of the Risk Value, the organisation can decide whether it can tolerate it or not. Now, all such risks are to be measured and monitored to assess and judge the decision.

Risk reduction: The hazard which cannot be tolerated will go for Risk Reduction to bring down the Risk Level to As Low As Reasonably Practicable (ALARP). The method of Risk Reduction is as follows:

- **Terminate**

  In this method, the risk needs to be terminated from route, such as replacing a high hazard chemical with a non-hazardous chemical. This method is no doubt very good, but application in practice may be difficult and so used in limited cases only.

- **Treat**

  This method is very common and very effective. Risk is treated by applying engineering control, administrative control, medical control, development of SOPs and implementation, detection system, PPE, training, etc.

- **Transfer**

  This method is allied to transfer the risk, such as insurance of company assets to cover the losses, or transferring the handling of hazardous materials, including processing and neutralising to government-approved agencies in their premises.

Implement and manage: The control measures finalised to reduce the risk is to be implemented systematically and retained on consistent basis.

Measure and monitor: Once the control measures have been implemented, it has to be measured and monitored. There is a saying that what can't be measured can't be controlled. Hence, the measuring criteria for control measures are to be finalised and the job should be monitored and recorded. The minor risk which falls under Tolerable category has also to be monitored.

Evaluate results/Investigate: The data obtained through measurement and monitoring need to be evaluated/investigated to judge the improvement in the risk level.

If the results show improvement in the Risk Level and falls under Acceptable Risk, this job shall join the list of risks for further risk value judgement. If there is no substantial improvement in the risk level, the risk will join the list of hazards and go for further risk analysis.

This is the way the risk cycle rotates and continuous reduction in risk level takes place, thereby ensuring continuous improvement in safety in the workplace.

Chapter 2
Operational safety

Safety is taken into consideration in a process industry right from the conceptual stage of design. Safety is thereby impregnated in design, layout, engineering, construction, pre-commissioning, commissioning, day-to-day operations and turnaround. This chapter gives an overall picture of the safety measures embraced at various stages in any Process Industry. I have added my personal experience also at various stages to give depth to the understanding and make it more meaningful.

Chapter 2
Operational safety

2.1 Operational safety

Keeping focus on productivity, safety, health, environment and reliability, running of process units and their allied facilities smoothly on a continuous basis is of prime importance.

Operational safety in a plant starts from the conceptual stage of plant design and ends with the day-to-day smooth, uninterrupted running of the plant in total safety.

1. The following tools are used to assess the level of in-built safety in the system:
   - **HAZOP:** Hazard Operability Study
   - **HAZID**: Hazard Identification
   - **PHA:** Process Hazard Analysis
   - **LOPA:** Layer of Protection Analysis
   - Other safety reviews such as **QRA**: Quantitative Risk Analysis, etc.

A. Hazard Operability Study (HAZOP)

The HAZOP study is to carefully review a process or operation in a structured and systematic manner to determine whether deviations from the design or operational intent can lead to undesirable consequences. Suitable guide words are used to create deviations in process parameters. This technique can be used for continuous or batch processes and can be adopted to evaluate written procedures. The HAZOP team creates a plan for the complete work process, identifying the individual steps or elements. This typically involves using the Piping and Instrument Diagrams (P&ID) or a plant model as a guide for examining every section and component of a process. For each element, the team identifies the planned operating parameters of the system at that point: flow rate, pressure, temperature, vibration, and so on. The HAZOP team lists potential causes and consequences of the deviation as well as existing safeguards protecting against the deviation. The team leader controls the discussion, so that meaningful results are obtained. When the team determines that inadequate safeguards exist for a credible deviation, it usually recommends what action be taken to reduce the risk in a comprehensive worksheet.

In the case study on FLIXIBOROUGH given in this book, it was a case of not doing proper HAZOP before putting a by-pass line to the reactor.

Objectives of carrying out a HAZOP study:

- To check a design
- To decide whether and where to build
- To decide whether to buy a piece of equipment
- To obtain a list of questions to put to a supplier
- To check running instructions
- To improve the safety of the existing facilities

HAZOP team:

In addition to the Chairman, the HAZOP team may comprise of the following personnel:

- Design consultant/project manager
- Production manager
- Chemical engineer
- Maintenance manager
- Electrical engineer
- Instrument engineer
- HSE engineer

B. Hazard identification (HAZID):

HAZID is a qualitative technique for the early identification of potential hazards and threats affecting people, the environment, assets or reputation. The major benefit of the HAZID study is to provide an essential input to project development decisions. It is a means of identifying and describing HSE hazards and threats at the earliest practicable stage of a development or venture.

Objectives of the study:

- To identify the potential hazards and to reduce the probability and consequences of an incident in the site that would have a detrimental impact on the personnel of the plant, properties and environment.

Scope of the study will depend on the particular project.

Methodology:

The study method is a combination of identification, analysis and brainstorming by the HAZID team members in a structured and systematic manner under a team leader, who controls the discussion so that meaningful results are obtained. Guidewords are used in order to identify possible potential and hazardous effects as well as threats. Furthermore, the team analyses the appropriate controls that should be put in place in order to prevent or control each identified threat.

The analysis of HAZID will be conducted on a session basis, grouping the processes with the Process Flow Diagram (PFD) and plant layout into a series of sections where the various sources will have similar characteristics and hence consequences. The entire discussion is to be recorded in a prescribed data sheet and submitted with report.

HAZID Team:

In addition to the chairman, the HAZID team will be constituted on the same pattern as in case of HAZOP.

Benefits of carrying out HAZID:

- Identify opportunities for inherent safety.
- Identify fire, explosion, toxic-release scenarios and measure to prevent it.
- Any special preparations required to be taken to handle these can be pre-planned.
- Any specific process modifications if required can be established at an early stage.
- Prepares the system and team, ready and confident to go ahead for commissioning. Avoids major surprises.
- Hazards involved in operating each equipment can be enlisted at the beginning, leading to better process mapping and better control in future for getting OSHAS/ISO approvals.
- The major benefit of HAZID is early identification, and assessment of the critical health, safety and environmental hazards provides an essential input to the project development decisions.

C. Process Hazard Analysis (PHA)

PHA is a systematic assessment of all potential hazards associated with an industrial process. It is necessary to analyse all potential causes and consequences of:

- Fires
- Explosions
- Releases of toxic, hazardous or flammable materials, etc.

Focus on anything that might impact the process, including:
- Equipment failures (refer case study No.27 in chapter No.3 of Flixborough disaster in 1974)
- Instrumentation failures or calibration issues
- Loss of utilities (power, cooling water, instrument air, etc.)
- Human errors or actions (refer Case Study No.-12 in chapter No.3 Bhopal Case study in Union Carbide in December 1984)
- External factors, such as storms or earthquakes, etc.

Both the frequency and severity of each process hazard must be analysed:
- *How often could it happen?* Tank spills could happen any time there, if there is a manual fill operation (multiple times a year).
- *How severe is the result?* Localised damage, fire, explosion, toxic gas release, death.

Core to the PHA analysis is the fact that things can and do go wrong. You have to forget IF it will happen and instead consider WHEN it will happen. Each identified hazard is assigned an 'acceptable' frequency. For purposes of the PHA, you cannot assume a hazard will 'never' happen.
- A hazard that results in simple first aid could be considered 'acceptable' if it could happen only once a year.
- An explosion and fire due to a tank rupture could have an 'acceptable' frequency of once in 10,000 years.

The end result of the PHA is a list of all possible process hazards with each one assigned an acceptable frequency of occurrence. The next step in the safety life-cycle is the layer of protection analysis.

Inadequate study and implementation of PSM led to many disasters in process industries. One of them being BP Refinery disaster, Texas.(please refer case study no 31 in chapter 3 for details.)

D. Layer of Protection Analysis (LOPA)
No single safety measure alone can eliminate risk. For this reason, an effective safety system must consist of protective layers. This way if one protection layer fails, successive layers will take the process to a safe state. As the

number of protection layers and their reliabilities increase, the safety of the overall process increases.

Some specific examples of protection layers include:

- Fire suppression systems
- Leak containment systems (dikes or double walls)
- Pressure relief valves
- Gas detection/warning systems

E. Quantitative Risk Analysis (QRA)

QRA is proven as a valuable management tool in assessing the overall safety performance of a chemical process industry.

Objectives of QRA:

- To identify, quantify and assess the risk from the facility, from the storage and handling of chemical products.
- To identify, quantify and assess the risk to nearby facilities/ installations.
- To suggest recommendations in order to reduce the risk to human life, assets, environment and business interruptions to as low as reasonably practicable.

Risk Analysis techniques provide advanced quantitative means to supplement other hazard identification, analysis, assessment, control and management methods to identify the potential for such incidents and to evaluate control strategies.

Risk Assessment procedure: Assessment of risks is based on the consequences and likelihood.

- Consequence estimation is the methodology used to determine the potential for damage or injury from specific incidents such as jet fire, BLEVE, etc.
- Likelihood assessment is the methodology used to estimate the frequency or probability of occurrence of an incident.

Software packages, such as PHAST RISK MICRO 6.7, WHAZAN v2.0, and EFFECTS v2.0, are used to carry out the modelling of probable outcomes such as fire, explosion, vapour cloud explosion and BLEVE.

Risks are quantified using this study and ranked accordingly based on their severity and probability. Acceptability of the estimated risk must then be judged based upon criteria appropriate to the particular situation. Study report is used to understand the significance of existing control measures and to follow the measures continuously. Wherever possible, additional risk control measures are to be adopted to reduce the risk levels.

2.2   Plant layout

Layout: To decide separation, isolation, drainage, accesses/roads considering foreseen scenarios for normal operation and emergencies. All regulations/ standards/statutory clearances shall be complied.

3-D review of plant during layout and engineering is an important tool to identify gap in safety and provides an opportunity to improve the safety standard in design stage.

While deciding the layout, the findings of QRA, population/facilities beyond factory area should also be taken into consideration. HSE aspect and best engineering practices shall also be taken into consideration.

The layout should also facilitate primary containment of chemicals in case of leakage/spillage for their safe recovery. Relevant statutes shall be followed as minimum requirement

2.3   Fire protection measures

In addition to fire prevention and protection measures in design, engineering and layout, it should be supplemented to take care, if it does occur. In the first phase, passive fire protection measures need to be considered, which may include fire-proofing of structures/ equipment, use of intrinsically safe/flame-proof electric equipment as per area classification, use of earthing/bonding in equipment/pipes, use of flammable gas detection system, etc.

The Active Fire Protection System may include well-designed emergency communications, fire detection system, firewater system, other automatic as well as manual firefighting systems, fire extinguishers at plant site. All such facilities and equipment should be freely accessible and well maintained for use during emergency. The facilities in premises may include a full-fledged fire service, stationed round the clock at fire station(s) with crew within the complex. The facility at the fire station shall include provision of mobile fire and rescue tenders with various firefighting facilities and chemicals.

Additionally, the complex should have well-developed 'Emergency Response and Disaster Management Plan' for all foreseen on-site as well as off-site emergencies. While preparing plan, the HAZOP and QRA findings are to be referred. The complex should identify all resources in terms of equipment/facilities and manpower as per requirement and should develop mobilisation plan. For the purposes, the complex may seek help from nearby industries (in form of Mutual Aid Scheme), government agencies, NGOs, medical services, etc. The requirement is mandatory in nature for most of the process industries in India.

An evacuation plan with Assembly Points should exist at all working locations for safe evacuation of people in case of emergency. This may be included as part of 'Emergency Response and Disaster Management Plan'.

2.4   Plant commissioning

Once the plant is ready, it has to go through various checks and documentations to ensure that plant is ready for safe commissioning. This includes

- All statutory permissions/clearances are in place and recommendations implemented.
- Plant mechanically ready and checked manually as per the prescribed checklist.
- All instrumentation system and electrical system and utilities facilities (such as power, air, nitrogen, water, steam, etc.) checked and are in place as per the prescribed checklist.
- All emergency handing facilities are in place and are in ready-to-use condition.
- Operating manual (including SOP, start-up and shut-down procedure, process emergency handling, utilities failures, etc.), maintenance manual, instrumentation manual, fire and safety manual, HSE management system, etc. are in place.
- Commissioning procedure has been established, and concerned persons are fully aware off the steps to be followed.
- Pre-commissioning audit has been carried out before the commissioning and recommendations have been implemented. In case it requires some statutory audits/checks, those need to be completed.
- Trained and experienced working personnel have taken their charge for safe and smooth commissioning of plant.
- Expert team from licensors, if needed, is available.
- Safety system addressing work permit system, PPE, emergency handling, MOC, etc. is in place.

2.5　Safety during plant operation

I. The additional document/procedure, which can strengthen the operational safety, includes:

(a) Competent and healthy person(s) should only be placed for field work. Pre-employment medical check and periodic medical checks are to be carried out as company's policy and/ or statutory requirement and records should be maintained.

(b) Starting for induction and familiarisation, training and re-training should be part of culture to be maintained by the company to keep their employees updated.

(c) Provision of log book for shift in-charges, instruction register, work permit register, incident register, PPE register, maintenance notification register, training register, process mock drill register, etc should be adequately made.

(d) Procedure for critical activities, such as furnace lighting procedure, caustic/water draining procedure from LPG vessels, sampling procedure, etc. should be displayed at site. Safety signs, hazard level and Material Safety Data Sheet (MSDS) for chemicals, important dos and don'ts should also be displayed at site.

(e) Interlock by-pass procedure and authority with alternate arrangement during by-pass should be in place. Any interlock by-pass has to be with approval of competent authority and to be recorded in instruction book for knowledge of the operation people as well displayed on boards beside of DCS panel.

(f) Procedure to record 'excursion of Safe Operating Limit (SOL)' and their investigation and monitoring. The procedure is commonly known as Process Safety Performance Indicators (PSPI).

(g) Procedure of Task Observations should be in place, and record of the same should be maintained. For the purpose, total task should be identified and critical task should be segregated. The frequency for observing each type of task should be decided. (Case study on over-flowing of kerosene from the bullet in a refinery is discussed in CHAPTER III.)

2.6   Training to personnel

The Operating personnel should be thoroughly trained on SOPs, emergency procedures and systems, action in case of utilities/equipment failure, safe start-up and shut-down procedure, handing over and taking over procedure, preparation of equipment/facilities for inspection and maintenance, etc.

Safety training should include SOPs, work permit system, Lock-Out and Tag-Out (LOTO) procedure, incident reporting procedure, management of change procedure, PPE requirements, emergency handling procedure, including hands-on exercise on fire fighting and PPE, safe maintenance procedure, etc.

Sufficient manpower should be exposed to first aid training to ensure that first aid trained person is available during operation of the plant. The list of authorised first aiders should be displayed near the first aid box in units/areas.

Knowing the procedure is not enough, unless it is implemented and followed in true spirit. Hence, the system should be in place to check compliance at site.

2.7   Inspection and Maintenance

Inspection, testing and maintenance of equipment and facilities shall be carried out as per the approved frequency. Preventive and predictive maintenance of equipment and facilities need to be carried out on time without fail.

Many companies practise Risk-Based Inspection (RBI) and Reliability-Centred Maintenance (RCM) for inspection and maintenance for equipment/facilities. If so, the same should be practised religiously.

Some of the equipment and facilities do need statutory tests, which may include lifting tools and tackles, pressure vessels, boilers, etc. The statutory requirement must be complied. Guidelines available in standards for frequency of unit turnaround and storage tanks, etc. shall be followed.

Pressure vessels also need periodic inspection and testing because of normal wear and potential corrosion either at welds or in the base material. The combination of pressure and volume determines the hazard: high-volume, low-pressure systems can have the same potential energy for release as low-volume, high-pressure systems.

When potentially corrosive chemicals are used (acids, caustics), or the plant atmosphere is corrosive (maybe near the ocean, or from chemical releases within the plant), what measures are taken to ensure the system integrity? Examples include periodic pressure testing, X-ray, etc.

Whenever a safety interlock of an equipment/facility is required to be by-passed due to one or other reason, this should be entered in interlock by-pass register and approval should be taken from competent authority based on criticality and/or duration of by-pass. The alternative safety measures should be categorically spelled out during the period of by-pass. It is a sound practice to display the same near the operating panel and entered in the instruction book. Once the interlock is attended and taken in line, this should be entered in the by-pass register.

2.8   Permit to Work (PTW) system

Procedure and compliance of the Work Permit System is the heart of safety in any process industry during inspection, maintenance, testing and construction activities. Policy of 'No Permit, No Work' should be in place.

This is a written document to carry out particular job safely, avoiding any communication gap.

Work permit in process industry is also a statutory requirement. Permits are issued in the prescribed format, depending on the type of work. Format has a checklist, which is to be addressed by permit issuer. There is also provision for the entry of gas test readings, PPE in the permit format.

Permit shall be issued by owner of area; like in process unit, it will be by operation officer and will be received by officer of the executing department such as concerned maintenance personnel for maintenance jobs. All permit signatories must be thoroughly trained and approved by authorised person for the purpose.

For jobs such as hot work, entry into confined space may require mandatory testing of gases. All such detectors should be timely calibrated and maintained in healthy condition. Persons, using these testers, should be thoroughly trained in use, advantage and limitations of equipment and interpretation of gas tester reading.

The type of work permit formats may include:

- Hot work permit/vehicle entry permit
- Confined space entry permit (case study in CHAPTER III on this point)
- Cold work permit
- Excavation permit

- Road cutting/blockade permit
- Electrical permit (energisation and de-energisation), including LOTO system
- Work at height permit
- Radiography permit, etc.

Job Safety Analysis (JSA) is a structured technique, used to identify risks at critical maintenance, inspection, testing and construction jobs, and take suitable measures to reduce the level of risk. Many companies practise it for all maintenance and construction-related jobs.

During performing JSA for a task, the task is broken into logical steps. At each step, hazards and risks are identified with control measures. After control measures, again risk level is assessed to ensure that risk has been reduced to an acceptable level. JSA is carried out by a team of operation, maintenance and safety officers; any other knowledgeable officer may be included as per requirement.

The JSA is used to supplement the safety steps of the PTW system. The findings of JSA are enclosed in the format to permit implementation at site.

2.9   Personal Protective Equipment (PPE)

PPE is considered as the last layer of protection against the risk. The minimum basic PPE requirement for anyone entering in plant/area should be defined first. Many companies define safety shoes, safety helmet and safety glass as the minimum PPE requirement for entry into battery area. This needs to be categorically displayed at entrance.

Some of the companies use the concept of 'WEAR 3 & CARRY 3' for PPE. **WEAR 3** means everyone in plant/area will always use three PPE (safety helmet, safety shoe and safety glass), and **CARRY 3** means everyone will always carry three PPE (safety gloves, respirator and ear protector) with them.

For persons working regularly, these PPEs are to be issued to them individually.

The premises should maintain respiratory-type as well as non-respiratory-type PPEs as per normal as well as emergency requirement.

Other PPEs are job-specific. The assessment of PPEs for various jobs should be made for the area, and the same should be available in unit/complex. The list of PPEs required for immediate job or for emergency handling should be readily available in process unit/areas. Such equipment should be checked/ tested as per schedule by HSE/operation personnel and the record should be maintained. List of equipment may include Self-Contained Breathing Apparatus (SCBA) set, escape set, PVC/neoprene suits, water gel blanket, fire proximity suit, ear protector, face shield, various types of respirators, furnace glasses, chemical suits, etc., depending on the type of hazards.

All PPEs shall be BS/EN-marked and shall be periodically inspected and maintained. The equipment shall be marked as 'out of service' on expiry or when it is damaged and shall be removed from the workplace with an immediate replacement. The standard guidelines and OEM recommendations shall be followed in discarding any PPE.

2.10   Management of Change (MOC)

All process modifications and changes in SOP should be subjected to Management of Change procedure. All such proposals shall be reviewed as per procedure by various departments, at various levels to assess the safety and operational requirements (HAZOP study in case of process modification), and on approval, changes shall be incorporated at site as well as in documents/ drawings. Training shall also be imparted to operating personnel on changes before implementing them.

The improper MOC during retrofitting has led to many accidents worldwide(please refer to case study no.27 in chapter3.)

The primary agency of the MOC system lies normally with the technical services departments.

2.11   Workers' participation in safety

The system shall be in place to involve all levels of employees in safety management system. Safety is a continuous journey and not a destination. The safety system can only be effective when all levels of employees, including contractor employees, are involved in system development and their implementation. The various forums used for the purpose include:

- Safety committee meetings
- Safety training and awareness programmes
- Safety competitions and appreciation
- Job Safety Analysis (JSA)
- Task observations

- HSE motivation and awareness campaigns
- Reporting of incident and sub-standard acts/practices/conditions
- Multidisciplinary safety audit (internal as well as external) and compliance of recommendations,
- Regular and statutory health check-ups.

2.12 Contractor safety

In today's scenario, many jobs and responsibilities are assigned to contractors. The contractor employees should be exposed to the same level of safety as company employees.

The contract document should specify their roles and responsibilities, skill and knowledge requirements and responsibilities with respect to safety and health. Only competent contract employees should be given charge at work site. Necessary training on safety, health and environment shall be given at the induction stage and later refreshers to keep them updated. Health check-up of all workers coming inside may be done, and fitness certificates are to be obtained. This may be done on an annual basis.

The contractor employees should be involved in all safety-related activities, to ensure their active participation in company's safety management system.

Safety appreciation and instruction shall go hand in hand in the work site. All positive behaviours shall be appreciated and recorded as part of company's programme to reinforce safe behaviour. All shortcomings observed during task observations and plant visit shall also be addressed to change the behaviour of the person.

Persons shall be encouraged to report substandard act/practices and conditions and shall be suitably rewarded for reporting. For this purpose, a 'Drop Box' shall be provided at work sites. All such reports shall be quickly addressed. High-potential observations (which may lead to severe injury/ fire/explosion, etc.) shall be investigated, and recommendations shall be implemented. All findings shall be shared with employees during safety committee meetings and other occasions.

Traffic safety will also be addressed to encourage safe driving habits. Habits of over-speeding, non-wearing of seat belts and carrying passengers on vehicles not registered as passenger's vehicle shall be discouraged.

The improper implementation of contractor safety have in the past led to fatal accidents of many contractor employees. Refer case study for accident during working on hydro-jetting and LPG unloading (case studies no.29 and 13 in chapter 3).

2.13 Good housekeeping

'**A place for everything and everything in its place**' is commonly known as good housekeeping. Maintaining good housekeeping not only leads to reduction of incident of fire and accident, it also keeps the morale of workers high. Examples of good housekeeping include:

- Floors and work platforms that are free from slip, trip, trap and fall hazards.
- Monkey ladder with provision of cage and barriers at exits; staircase with hand rail and toe board.
- Garbage/ dustbins for various types of materials and an organised system of garbage collection and disposal.
- Lube oil and other drums at their specified location. Measures in place to collect spillages and wastage and their disposal.
- Practice to remove surplus materials/scraps/ debris materials from work site on completion of work and the same properly implemented.
- Foundation of equipment (pump/compressors, etc.) kept free from oil/ deposits and scraps.
- Adequate number of toilets/washrooms/wash basins/check and change rooms are available and maintained in hygienic condition for all.
- PPEs are maintained properly and kept at specified locations.
- All pipelines are painted as per their colour code. Direction of flow are marked on pipelines.
- All safety signs and procedure for critical activities are displayed.
- All emergency communication facilities and equipment are readily accessible and well maintained.

2.14 Mock drills

In addition to process mock drills, the plant should also carry out emergency drills at specific intervals. The drill should include evacuation drill and casualty handling. Depending on the level of emergency, the mutual aid partners should also be engaged time to time in drill. Any gap observed should be noted and corrected timely

2.15   Safety inspections and audits

Periodic safety inspection/audits shall be carried out by multidisciplinary team; normally, it is conducted annually in line with various statutory requirements. The recommendations of the audits should be implemented in a time-bound manner.

2.16   Safety during shut-down

Plant can go to shut-down in a planned manner for scheduled maintenance/ idling or due to some emergency. Normally in shut-down, many external agencies are involved and many workers of various skill levels are deployed. It is a difficult task to bring the same safety level to a functioning work site.

In a planned shut-down, since it is a planned activity with known jobs, the safety activity during the shut-down should also be planned in a manner that is incident-free. The safety management system should be incorporated in the contractor document, and the contractor should be held clearly on safety commitment. Only a contractor with a good safety record should be allowed to participate in the contract. For critical jobs, specialised agencies may be lined up. The list of jobs may include scaffold erection and dismantling, equipment erection, catalyst replacement, material handling (crane, fork lifts, excavators, JCBs, etc.), hydro-jetting/sort blasting, etc. The main contractor shall have his own safety officers in the set-up to ensure safety. At site, the system may focus on:

- Mandatory safety training for all persons coming for shut-down,
- Tool Box Talk (TBT) before the start of work in each shift by officer from executing department. In some cases, the contractor may also be allowed for TBT.
- Specialised training for critical activities, such as hot work, entry into a confined space, hydro-jetting/shot blasting/radiography, work at height, etc., shall be arranged by the safety department. Also, on system of PTW, JSA training shall be arranged.
- Entry of duly tested lifting machines, lifting tools and tackles shall only be allowed at complex, and the record shall be closely monitored. These equipments shall be subjected to physical check for safety before entry in the premises by the mechanical department.

- To ensure the use of only standard PPEs, the contractors shall submit a sample each of PPEs used to company's safety officers for approval. The approved type shall only be used at site by the contractor.
- All equipment erection shall be carried out as per the approved erection plan. Compliance of PTW and JSA shall be the backbone of safety at the work site.
- Safety monitoring shall be intensified at sites to ensure safety compliance.
- All other equipment used at site, such as hand tools, welding machine, cutting and drilling machine, shall be checked and certify for their fitness before being used. Such certificate may be valid for one month.
- Duly certified scaffolds with GREEN TAG only shall be used for the work.
- The safety department of complex shall coordinate closely with safety officers of contractors to improve safety compliance at work sites.

In case of emergency shut-downs, the planning may not be possible in advance. The existing safety system shall be followed for carrying out the job safely.

2.17   Plant start-up after shut-downs

On completion of shut-down after boxing up of all equipment and removal of surplus materials/scraps/debris, closing all the work permits, the 'Pre-Start-up Safety Review (PSSR)' is to be carried out by all departments in a prescribed format. Departments need to certify that all planned jobs have been carried out and the equipment is checked and found ready for start-up. For proper line-up of pipeline and equipment, refer drawing/procedure to ascertain, as any unnoticed blind in the system can cause a major accident, or vice versa.

Safety department will also carry out safety checks as per checklist, which will necessarily preclude all fire protection facilities and safety system be in place, that all permits are closed, that the unit is free from all surplus materials, scraps and debris and all temporary facilities for shut-down. These include the removal of all mechanical and electrical equipment, contractors shed, their manpower from unit battery area/hazardous zone. All observations by various departments need to be addressed by plant manager, before allowing the plant to go for start-up.

This procedure may not be followed, when the unit is being started up after idling but where no maintenance job has been carried out during shutdown. Not implementing the PSSR had led to disaster in the BP refinery, Texas, during start-up of ISOM unit in March 2005. (Refer case study on disaster.)

There are many cases of tray dislodging in distillation columns due to water ingress in feed, poisoning of catalyst due to sulphur slips, fires due to flange leaks or thermal shocks and coke formation in heater tubes due to overheating, even leading to heater fire.

Start-up of a process unit after major turn-around, is a very critical activity for smooth operation and safety of plant. All the persons attached to the plant have to be very alert in following the procedure for line-up, replacing of air from system, raising the system temperature slowly, keeping constant watch on any leak, skin temperature of heater tubes, abnormal rising of pressure in any vessel, establishment of flows as expected. Observing and continuous logging all these parameters are very important and the in-charge needs to keep an eye on these observations. Any abnormalities noticed at this stage need to be rechecked and confirmed, and corrective measures must be taken immediately.

Chapter 3
Case studies

There are many case studies available, but we have selected only a few to demonstrate the types of accidents in industries. We have analysed their root causes, and the manner in which they occur? The questions we have asked ourselves are: why and how do these happen? And what are the consequences: what are the damages and losses? We hope the readers will get a fair idea from these case studies and on their own, research more and more such case studies easily available on industry websites and circulated by safety departments in their bulletins in various organisations.

Chapter 3
Case studies

Case studies

In the following pages, I would like to narrate some incidents and my learnings from them. These incidents are from my own collection, incidents in public domain, and the reviews are what I made during my forty years' working experience. I have given my analyses and recommendations, but these should be treated only as broad guidelines. They may have to be revised on case-to-case based on the actual situation and type of industry.

3.1   Pump house fire

I had just completed my morning round of the unit and reached office when suddenly, I heard the fire siren. The phone rang urgently to inform me about a fire in the atmosphere unit (CDU). On reaching the site, I found that the residue pump was on fire and the fire-fighting team was trying to extinguish it.

Reason: Atmospheric residue pump is pumping hot residue from column bottom to run down, as the product (around 300 °C) is at more than its auto-ignition temperature. As soon as there was a leak from the flange, it caught fire.

Learning: The pump was under commission after maintenance and the flange started leaking. It seemed that the gasket used in the flange joint was not suitable for such hot service. Hence, after the temperature was increased, the gasket failed, leading to the fire.

We need to be very careful in selecting the right/good-quality gasket. A small detail like this can lead to huge production loss, as the unit has to shutdown for a whole day.

3.2   Fire at crude oil heater due to leakage in outlet flange

A leak from flange joint in transfer line from the Heater to the column was noticed. The crude oil in the coils was replaced by diesel, he transfer line was opened and the heater was left for cooling. Suddenly, there was fire at the outlet flange as the diesel inside got heated up (even though firing was cut off, the box was hot) and reached its auto-ignition temperature. The fire increased as the operator opened the steam into the coil (a normal procedure in case of furnace fire due to leak in coil, hydrocarbons are pushed to the column by steam).

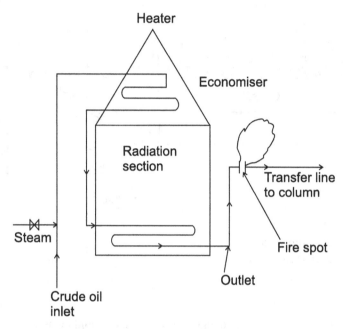

Reason: By exposing the coil to steam, the diesel in the coils got pushed out to the atmosphere and this aggravated the situation. The operator should have known that in this particular case, he was only pushing the hot diesel out and hence doing the wrong operation.

Learning: One should take a moment to analyse the situation fully and only then take action. The operator should have taken into consideration the fact that if there is a fire in the heater due to tube leak, the oil is pushed out by steam. Blindly following normal procedure, without factoring in developments can, and often will, aggravate the problem.

3.3   Avoid shortcut

A technician working with a construction contractor started walking from the fabrication area, having decided to take his lunch in the shade along the boundary wall of the refinery. When he reached point A, he thought to take a shortcut and cross the drain by walking on the water pipe line instead of going over the bridge. While crossing in this manner, he lost his balance midway and fell. His head struck the sidewall of the drain. He was taken to the hospital but was declared brought dead.

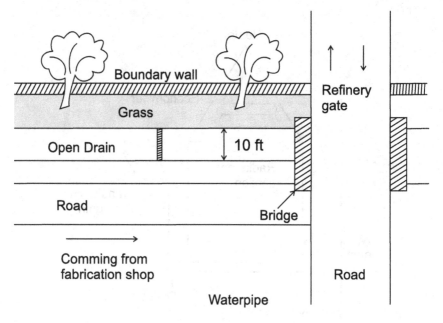

Learning: Shortcuts are dangerous and hence should be avoided.

3.4   Never allow anyone under the influence of alcohol inside the premises

I have come across two such cases where the operator under the influence of alcohol came on duty and met with an accident.

Incident A: Operator X was a habitual drinker. One night he came on duty. He was working in the refinery's oil movement and storage area. The shift in-charge sent him to start the pump in order to load diesel into tank wagons. While walking to the pump house, in his stupor he did not notice the danger board which had been placed in front of an open manhole, kept ready for cleaning, and fell into it. Since that area was deserted, no one could hear him calling for help and he lost consciousness. After a while, the shift in-charge came searching for him and found him lying in the manhole. Fire-fighting crew and ambulance was called. He was taken to the hospital but due to inhaling too much hydrocarbon vapours in the manhole, he could not survive.

Incident B: A similar incident took place in another petrochemical plant, when an operator in drunk condition climbed a tank wagon for loading the product but lost his balance and fell down on the ground. He succumbed to head injury in the hospital.

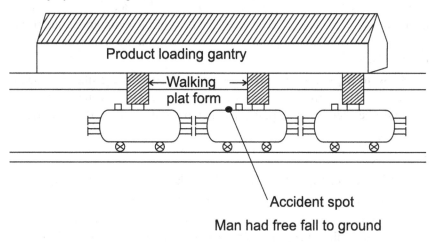

Accident spot
Man had free fall to ground

Reason: Under the influence of alcohol, a person's reflexes are not good and the body balance is not under his control. Hence, making mistakes under such circumstances is very normal.

Learning: A person under influence of alcohol can put his life and the lives of his colleagues and the plant area in danger. Such persons should not be allowed inside the factory/industry premises. In many locations, it is normal practice to carry out the breath alcohol test for everyone at the entrance. When I was CEO in Mombasa Refinery in Kenya, even I had to go through this test while entering the refinery.

Also, there should have been a proper barricading around the manhole to avoid any such incident of person falling inside.

3.5   Kerosene over flow due to no supervision

Incident: In one industry, one sulphur dioxide bullet came back after maintenance. After a hydro-test, the water was drained out. In order to make the bullet completely moisture free, kerosene was being filled in it. A single operator was supervising the operation. He remembered some urgent job in the bank outside the premises and thought he would be able to be back in time and left the site unattended. After completing the bank job, he went to the canteen for lunch. In the meantime, the bullet was full and kerosene started overflowing. It found its way into an open drain. Fortunately, the fire and safety group came for a routine round and noticed the hydrocarbon smell along the open drain. Immediately they cordoned off the area, brought the foam tender and spread foam over the top layer of the water level/kerosene to avoid fire. With the help of sand bags, the flow of the drain going out of the complex was stemmed. After great effort, the kerosene was removed from the drain and a major fire averted.

Reason: Negligence on part of the operator could have resulted in a major emergency. If there was some urgency for him to leave the site, he should have closed the kerosene inlet valve to the bullet to avoid overflow. It was a case of gross negligence.

Learnings:
- In the work permit/SOP, there should be a condition for a person to be continuously present at the site for stringent supervision of the job.
- The storm water drain connection to open drain should be closed while doing such operation to avoid the flow of oil/chemicals into an open drain.

3.6 Cleaner sleeping under a truck

Each industry has truck loading or unloading facility in the premises. Safety is a big challenge while regulating the movement of so many trucks in a day. Training of drivers and cleaners for safe operation is one of the essential parts to maintain safety; still, accidents happen due to some negligence or the other. It is a fact that most accidents have occurred while trucks are reversed.

Incident: One afternoon, when the truck loading operation was stopped for shift changeover, one cleaner decided to take rest for a while under the parked truck. In no time, he fell asleep. When he operations resumed in the second shift, the driver started the truck and reversed it to align it to the loading bay. He was not aware of the person sleeping underneath the truck; as a result, the cleaner was crushed under the wheels and died instantaneously.

Learnings:
- There should be a proper rest room for drivers and cleaners/helper.

- Driver should always take the help of cleaner/helper before reversing the truck.
- Cleaner/helper should guide the driver from rear side while reversing the truck.

3.7   Risks in crane operation

In day-to-day operations in all industries, sometimes accidents take place which are difficult to envisage, despite precautions already taken.

Incident: During one of the oil unit's shutdown, one chemical dosing pump was to be removed as pre-start-up activities were in progress. This was a pulley-driven pump. The boom of the crane lifted the pump to a great height, and before that, the area around the crane had been fortunately cordoned off for safety reasons. Suddenly, due to the heavy pull of rope on the cast iron pulley, the pulley broke into several pieces, and these sharp-edge pieces fell down.. One such piece fell on a worker who happened to be on the reactor platform. This piece managed to pierce through his helmet but could not do much harm. THE HARD HAT SAFETY HELMET HAD SAVED HIS LIFE.

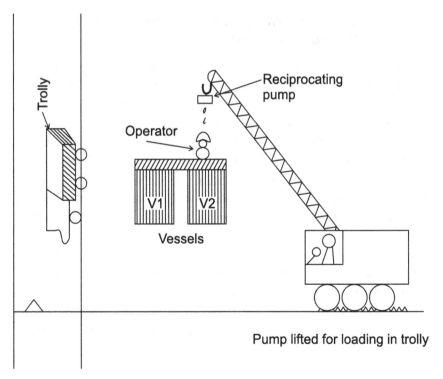

Pump lifted for loading in trolly

Learnings: The incident indicates that mere cordoning off the area at ground level is not adequate. We need to ensure that no one is present anywhere, at any height, t within the reach of the crane boom.

- Some warning system should be used before starting such an operation.
- While tying the rope around any equipment to be lifted, ensure that this is not putting pressure on some breakable part.
- The worker's life could be saved because he was wearing a safety helmet; hence, it is imperative to wear hard hat helmets in the plant area.
- The sling of the crane should be checked for its condition and load test.
- There are cases where the boom of the crane itself fails so the worthiness of the crane should be checked before its use.
- The crane should always be parked with boom in lowered condition and brakes properly applied.

3.8   Static electricity causing fire due to and improper earthing

One evening, there was an explosion inside a refinery. The fire siren went off ed and fire tenders rushed to the spot. Fire had started in a tank lorry loading motor spirit. The driver and the cleaner had died on the spot.

Incident: The truck had been parked in the loading Gantry or tanker loading platform to the load motor spirit. As soon as the cleaner opened the valve to load the truck, the spirit inside the truck caught fire and exploded.

Learning: Investigation revealed that the cleaner had forgotten to connect the earthing clamps to the body of the truck. Hence, as soon as the liquid motor spirit started flowing, the generated static electricity could not flow down to earth in the absence of an earthing connection.

Static electricity is an invisible source of explosion and fire, and hence, everyone has to be extra careful in ensuring that any vessel and equipment receiving hydrocarbons is properly earthed.

It is the job of the supervisor at the loading Gantry who must ensure that proper earthing connections have been made before giving clearance for loading.

3.9   Presence of mind can avert a crisis

Incident: In a unit handling liquid sulphur dioxide, a leak developed at a point upstream of its pump discharge valve, and this toxic gas started spreading rapidly. If not controlled in time, the whole area around the unit and its neighbourhood would have been affected. Operators were afraid of entering the gas cloud but one brave operator took up the challenge. He put on the PPE equipped with air breathing apparatus, rushed to the pump, closed its suction and discharge valve and put it off. This combined presence of mind, courage and knowledge prevented a dangerous situation from developing.

Learning: We should know how to use various PPEs available in the plant. This learning can save both the plant and human lives.

Training in proper use of PPEs is required regularly, and field staff should be confident of using these in cases of emergency.

3.10   Safety culture is driven from the top

As I mentioned in my introduction, safety is everyone's responsibility but initiative and push are required from top level. People generally try to avoid wearing basic safety gear like safety shoes and helmets. When this is allowed, chances of accidents are higher and people working in such organisations are not safe. However, once the management decides to impose discipline, the situation can be improved in no time at all.

I know one industry where when the new CEO joined, he saw employees inside the battery area coming in chappals and moving around the plant without their hard hats. He wanted to issue a circular proclaiming that strict action would be taken against those who did not wear safety shoes and helmets. The production chief was scared that the Union might not like this circular and agitate against it.

The CEO called a meeting of Union and Officer's Association representatives and discussed the matter. He explained that safety shoes and helmets are given to employees free of cost for the sake of their safety and that of the plant. Everyone had to use them while on duty and he needed their co-operation in creating safe working conditions. All agreed to the CEO's proposal. A notice was issued with a warning that after a month, if anyone was caught without safety shoes and helmet in the field, he would be fined. Gradually, the percentage of compliance started increasing, and within two months, 98% compliance was achieved.

At the construction site, a number of workers were observed to be working without safety shoes. The CEO ordered that safety shoes from stock lying in stores be issued to the contractual labour free of cost. Hard hats were also issued free on a one-time basis. Any replacement was to be done on the payment.

The CEO started taking morning rounds to shop-floor level to address the employees about safety, productivity and punctuality. The efforts brought a significant improvement in the attitude of the employees towards these aspects. At the farewell of their CEO, one and all appreciated his dedication to the safety and well-being of the employees and their families.

3.11   All petroleum product tanks on fire

This is the case of a marketing terminal storing products like motor spirit, diesel, ATF/kerosene. This was a well-designed terminal with a lot of safety measures and interlocks in place. The incident that happened was very unexpected and difficult to envisage even at the design stage and in the HAZOP study.

Incident: At around 5 pm one day, two operators were given the job of lining up one full tank of motor spirit to be pumped to the customer's tank, which was connected by a pipeline. However, one of them decided to first go to the canteen, and other operator entered the tank dyke alone for what is technically known as the de-blinding operation. The horror that followed as this: as soon as this operator loosened the nut bolts of the hammer blind to reverse the position, and line up the tank outlet to pump inlet, the motor spirit gushed out like a fountain and drenched the operator. The operator tried to call the control room to shut the body valve of the tank (electrically operated valve), but the engineer on duty had gone to witness the tank dip along with the customer's representative. Within no time, the operator became unconscious due to suffocation by motor spirit vapours.

The flow of motor spirit from the hammer blind continued uncontrolled creating a vapour cloud. The second operator came back from the canteen and tried to remove the unconscious operator from the site. But by this time, the whole tank farm was flooded with so much motor spirit liquid and vapours that he also suffocated and fell down unconscious in the dyke area. As time passed, the pool of liquid motor spirit kept on building up in the dyke area and an unconfined vapour cloud started spreading in the terminal. It being evening and most people off duty, no one was around to notice this alarming condition. The Production Engineer, meanwhile returned after the tank dip, and noticing the strong hydrocarbon smell, arrived at the site. He also started feeling nauseous and rushed to the hospital. By now, the message had reached senior officers but when they reached the terminal gate, they found it was impossible to enter, as the whole area was filled with hydrocarbon vapours.

It was only a matter of time before the vapours were ignited s and caused an explosion. resulted in huge explosion and fire. The whole terminal was so damaged that even the fire-fighting facilities became redundant. Soon,, all petroleum product tanks were on fire. The only option was to keep the product tanks cool by spraying water on their external surfaces and let the product burn inside, in a controlled condition to avoid tanks exploding and spreading the fire even further.

There were a number of casualties in this accident and enormous loss of products. The terminal was totally destroyed.

Cause and contributing factors:

- The root cause of the incident was negligence by operator who did not ensure complete closing of tank body valve before opening the hammer blind. Since the tank was full, the high liquid head in tank led to profuse flow of MS from the tank into the dyke area.
- The operating personnel who tried to control the leak were not wearing adequate PPEs.
- All operating personnel were overcome by MS vapour. There was no one to control the situation.
- The MS leak continued and the vapour cloud spread further. It could have got ignited by ordinary light fittings in the administrative building or by even a vehicle. This led to massive unconfined vapour cloud explosion.
- Entire terminals,, was on fire, and there were multiple casualties. Decision was taken to allow the product in the tank to burn till the entire product s burnt out. The pipe connections of the fire-fighting facility were badly mutilated due to the explosion.

Learnings/Recommendations:

- Hammer blind valve in the tank outlet should be replaced by Double Ball Bleed Valve (DBBV). The valve should be located outside the dyke. In this arrangement, no reversal of blind is required for line up.

- The first body valve of the tank should be the Remote-Operated Shut-Off Valve (ROSOV). It should be fail-safe and fire-safe type, and operated from control room, The operating switch should be located outside the tank dyke.
- A Radar gauge should be provided in each Class-A petroleum tank in addition to the existing positive displacement-level indicator/control. High-level alarm from radar gauge and a high-level alarm from separate tap should be installed.
- Hydrocarbon detector near potential leak sources for Class-A and Class-B petroleum products such as tank dyke, tank manifold, and pump manifold should be provided.
- Area should be covered by CCTV camera.
- Site-specific Standard Operating Procedure (SOP) should be prepared and implemented.
- Availability of PPEs (including self-contained breathing apparatus and fire suit) should be ensured, and training on use should be given. Each terminal should have emergency kits with necessary emergency handling items.
- Shift manning should be maintained as per the schedule.
- Extensive fire and safety training to employees, regular contractual employees and security personnel should be given.
- Rim seal fire protection system should be provided in each Class-A petroleum tank.
- Tanks should be protected with high-volume long-range foam monitor (variable type) for fighting tank fires.
- The fire water facilities with adjoining petroleum installations should be interconnected to improve reliability.
- Medium expansion foam generator should be provided on tank dyke to suppress vapour in case of any spillage.
- Internal safety audit should be strengthened for meaningful findings.

3.12   Escape of toxic vapours from chemical plant

(This case in India is well known, though the name of company and the place of incident are undisclosed here.) It was one of the major disasters of its kind where thousands of people died or became disabled. The incident showed that how a minor mistake by an operator can play havoc.

This chemical factory was running for a number of years and producing chemicals to be used in agricultural sector. The intermediate product known as Methyl-Iso-Cyanite (MIC) is highly lethal in nature. It was stored in a bullet of around 80 KL L capacity. There was regular monitoring of temperature and pressure.

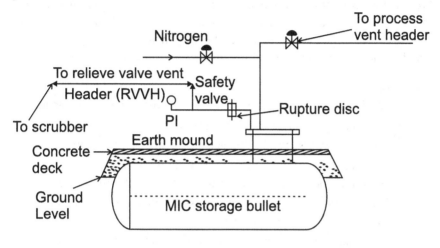

Simplefied sketch of MIC storage tank

The pressure relief system consisted of a rupture disc before Pressure Safety Valve (PSV) to ensure that PSV was not exposed to MIC vapours all the time. Vapours coming out of the PSV discharge were routed to Relief Valve Vent Header (RVVH). The vented material was routed to vent gas scrubber for neutralisation/detoxification. After neutralisation, the gas from scrubber was routed either to a stack for releasing to atmosphere or to a flare header to burn there.

The MIC tank, which was idle, had around 40 tonnes of MIC. Everything seemed to be normal during the day. During mid-night, the MIC vapour escaped the plant un-neutralised. This chemical is very toxic and can will affect everyone in the surrounding area. As a result, many any people in near by areas died in their sleep itself, and many became disabled/seriously ill.

It seems that some operators by mistake connected a water hose to the idle bullet. Water reacted with MIC, and due to exothermic reaction, MIC vapourised. This resulted in pressure built-up in bullet, leading to the rupture of disc and opening of PSV. The vapours travelled to the scrubber, but since the rate of vapour generation was very much more than what the system could handle, MIC (untreated) escaped into the atmosphere. Unfortunately, the flare was also under maintenance on that day.

Cause of incident:

From the investigations, only apparent cause of incident appeared to be the introduction of water in MIC tank by some operator without knowing the gravity of the damage it was to cause. However, had the flare been operating that day, the MIC would have burnt safely.

Learnings:

- Storing of minimum quantity of toxic chemicals inside the plant.
- Training people in the plant about the dos and don'ts.
- Carry out process 'safety management study' and HAZOP.
- Disaster management plan should be known to key operating personnel and conduct mock drill for on-site and off-site disasters.
- The production units dealing with such lethal chemicals should be located away from inhabited areas.
- In off-site disaster management plan, a support system from local administration, police and health authorities, should be included. diagram.

3.13   Fire in LPG truck loading/unloading facility

Incident: There was a heavy leak of LPG from a loaded LPG truck in a LPG bottling plant, followed by vapour cloud formation. The unconfined vapour exploded after coming into contact with a source of ignition, leading to a major fire in a LPG Tank Truck Decantation Facility.

Reason: On investigation, it was revealed that one truck was unloading LPG in a LPG bottling plant with proper flange connection at unloading point. Due to a push from another truck which happened to be reversing, there was tension on the flange connection. This caused a profuse leak of LPG, and the whole area was engulfed by LPG vapours in no time. The vapours were ignited from the DG set in the premises and resulted in a major fire.

Learnings:

- There should be enough space between the adjacent loading/unloading points so that any operation going on at the adjacent point does not cause any push–pull effect on the operating point.

- The DG set should be located at a safe distance, besides ensuring it has a spark-free operation.
- Automatic fire-fighting system with deluge valve sprinklers should be installed. These should be equipped with a gas-sensing signal. This should operate immediately, and spray water on the facility to protect it from severe damage.
- The operation and movement of trucks should be regulated at a safe distance, without disrupting or disturbing other activities.
- Whenever we are handling LPG in open areas like LPG bottling sheds and LPG storage areas, we must ensure that there are no cable/pipe trenches in the area as LPG vapours are heavier than air. These settle in lowlying areas and can be a cause of big fires sparked by some ignition source. If such trenches do exist, fill them with sand.
- As compared to LPG, natural gas (CNG/PNG) is much lighter than air, and in case of any leak from the system, it gets dispersed in the atmosphere. So, it is safer than LPG.
- We have come across many instances of leaks in LPG road tankers. Ministry of Petroleum and Natural Gas has given clear instructions to Oil Marketing Companies to train their TT crews how to handle minor leaks. In case any help is required, they should contact the nearest refinery or LPG bottling plant of any company, to control the leak and safely transfer the LPG to another tanker.
- Ensure adherence of SOPs for unloading operations by operators and drivers.
- Breakaway couplings of TLD arms installed in all LPG plants must be tested for efficacy in both forward and reverse directions of the movement of the bulk tank trucks.

3.14 Fatal accident due to asphyxiation inside a vessel/confined space

These types of accidents usually happen in a running refinery/chemical industry when during shut-down the vessels go for internal cleaning. One such accident took place in one of the refineries where I worked.

Incident:

The reflux drum of distillation unit had been given for cleaning. The drum was steamed and left open for cooling down and entry of air.. The blinds were inserted on the flanges connected to the column, outlet nozzle, etc. After tool room talk on the precautions to be taken, contractual labour was permitted to enter the vessel by the Production Department, after due work permits were issued.

The first operator was found unconscious inside the vessel within a few minutes of his entry. Anther operator went inside to rescue the first operator but soon he also collapsed. The third operator smelt e hydrogen sulphide gas inside the vessel and raised an alarm. The two operators were brought out and rushed to hospital but within a few days both lost their lives. A committee was set up to probe and analyse the reasons, which were as under:

Reasons:

1. It was a clear case of asphyxiation inside the vessel (non-availability of sufficient oxygen in the vessel). There was one fuel gas connection of one inch diameter connected to the vessel, which was suspected to be in unblinded position at the time of accident through which fuel gas containing hydrogen sulphide gas could have entered the vessel and since this gas is toxic in nature, could have caused the operators to lose consciousness.

2. The other possibility was the entry of nitrogen through a hose (mistaken for an air hose) kept for maintaining the fresh air supply inside the vessel as both air hose and nitrogen hose were very similar. Both were connected to pipes which were unmarked, thereby giving rise to confusion about their services.

3. The operators did not use proper breathing apparatus while entering the vessel.

4. Before entering the vessel, Production Department should have checked the oxygen percentage inside the vessel, which should ideally be above 19 percent.. Using Gas Meter, the absence of H_2S, etc. should also have been ensured for safe entry into vessel and supported by fresh air supply through air hose.

Learnings:

- Before issuing a work permit, operation group should identify
- hazards in carrying out the cleaning activity of the vessel from inside.

The operation group should clearly show the blind position in the vessel drawing, and ensure that the blinds have been inserted at right positions. This should be re-confirmed before giving the entry permit.

Such jobs should be carried out under supervision. One person should be deputed to stay outside. The person who enters the vessel first should have some whistle or a bell to alarm the person standing outside, so that an immediate rescue operation can be arranged.

There should be clear colour codes to demarcate differences in hoses used for oxygen and nitrogen and proper display boards to be put up at the tap-off points for easier identification of the service.

- It is very critical to ensure that the vessel is free of hydrocarbon and any other gas. The oxygen content in the vessel should be more than 19 percent.
- Suitable respiratory apparatus should be used before entering the confined space.
- SOP should be kept ready for such operations, and one should ensure that the procedure is being followed religiously.

3.15   Fire due to electric short circuit

Fires due to short circuit are common in countries with hot weather. In India, summers are very hot and, in certain regions, the air can also be rather dry. Continuous use of electrical appliances leads to overloading often,and the electrical wiring gets is heated up. The high ambient temperature adds to this problem. Insulation weakens, short circuiting takes place, and fire results. If such fire breakouts are noticed immediately, and prompt action taken, the fire can be controlled by using CO_2 extinguishers and by cutting off the supply from the mains.

Normally, when such short circuiting takes place at night time or on holidays, it results in big fires in offices, shops, warehouses and substations, etc. Very frequently, we hear such news on TV and read in the press about such fires, leading to a number of fatalities and huge damage to property.

Learnings:

- Only good quality wiring with ISI mark should be used.
- The rating of the cables/wire should be selected based on the highest load expected, and keeping some safety margin.
- CO_2 extinguishers should be placed at easily accessible places, and people should be trained to use them.
- In case of fire, the electricity connection is to be cut from mains immediately.

- No water or foam should be used to fight electrical fire.
- Time-to-time checking of the condition of cables is advisable to prevent such incidents.
- In closed system, there should be some ventilation provided to facilitate the dissipation of heat on a continuous basis.
- Overloading of the system is always dangerous and should be avoided.

3.16   Accident due to wrong line-up

Correct line-up for commissioning a process unit is most critical. It can make all the difference in the prevention of a prospective shut-down/fire/explosion. During start-up activities, a normal line-up is done. This is basically achieved by removing the blinds which were inserted for isolation, often reversing the blind position at many locations, dropping or replacing the spool piece into its original position, etc. I have seen cases where the unit had to be shut down again due to one blind not being removed by oversight.

Incident:

The atmosphere unit was in start-up mode. After the displacement of the air by steaming and fuelling gas back-up in the system, crude oil was received in the main column. Through cold circulation, the water-draining exercise was completed, after which, burners were put on to raise the temperature. Slowly, as the temperature went up, the product started coming in the side draw-off vessels, beginning with naphtha, kerosene and diesel. Once enough level was built up in the diesel vessel, the pump was started to send it to the run-down tank. Diesel was coming to the pump at more than 200 °C, pump pressure was increased to its shut-off pressure. One blind on the discharge side had not been removed. Suddenly, the gasket that follows the discharge valve gave way, and there was a sudden splashing out of hot diesel, followed by a fire. Immediately, the unit was shut down, and the fire brought under control. The system had to be cooled down again, the gasket was replaced, and the culprit blind was removed. Only then the unit could be restarted safely. This shows how a small slip in line-up can cause a hazardous situation and loss of productivity.

Learning: In the above case, there was only a small fire, which could be controlled immediately, but there was a loss of two man-days.

I remember another similar incident where there was explosion and fire due to some mistake in line- up before start up. In most of the Hydrocarbon and chemical industries, it is very crucial to make the entire system air free and water free before start up. The air left in the system can form an explosive mixture when the oil is heated and produces light fractions of hydrocarbons. Similarly, small amount of water, on heating can convert to steam and increase the system pressure in no time.

This incident happened in one of the refineries when FCC unit was ready to start. All the start up activities were going normal when all of a sudden there was explosion in the slurry settler vessel. The explosion was so strong that parts of this vessel flew away and hit the piping and vessels. The impact was so strong that it sheared the connecting pipelines and punctured some vessels. There were fires at no. Of places and the oil and gas coming out from these openings aggravated the situation to great extent. The structures and vessel supports were badly damaged and it took many days to re-built the unit and make it operational.

The Enquiry committee came to the conclusion that the explosion in slurry settler took place due to presence of some water in the vessel bottom which either could not be completely drained due to chocked bottom drain or condensate accumulated de to entry of some steam through steam connection, specially provided for de- chocking of the bottom drains. With heating of the oil in vessel to around 300 degree centigrade, water got evaporated suddenly and built up excessive pressure (many times more than design pressure) resulting in to rupture of this vessel. Delay in evaporation of water must have been due heavy and viscous oil leading to slow rate of heating of water lying in some corner of vessel bottom and as the temperature increased it evaporated in a surge.

The above study clearly explain the importance of Standard Operating Procedures, special attention and care of all during start up and training of staff to cater to emergency situations.

3.17   Fire due to line vibration

Sometimes we treat the situation lightly, not realising the repercussions of negligence. In one refinery, there were two hot naphtha pumps. The normal procedure was to operate one at a time. It so happened that operational pump 'A',, was shut down due to some vibration problem. Pump 'B' was then put into operation, but it also started tripping after just one day. To keep the unit running, Pump 'A' was put into service despite the vibration issue. There was a half-inch diameter pipeline connecting from discharge to suction side of pump with a needle valve in between. Pump 'A' continued to operate for 24 hours, then due to fatigue, the joint of this half-inch diameter line gave a way, releasing hot naphtha vapours, which then caught fire, and the whole pump house was on fire. This resulted in damage to a large number of pumps, motors, electrical and instrument cables. The unit was shut down for a week for repairs. So, in industry, we should not overlook the indication of any weakness in the system as it can result in major accident, loss of life and financial losses.

3.18 Accident during excavation

Incident: Outside one refinery, excavation was being done in order to lay a product pipeline. In the work permit, it was clearly stipulated that the earth removed was not to be dumped at the edge of the dug area and that there was to be proper barricading by tin sheets and bamboos, so as to prevent the earth from caving in. Unfortunately, due to lack of supervision and in order to save time, the labourers did not follow any of these instructions. When the pit became about two metre deep, all of sudden the wall caved in and the whole heap of loose earth fell into the pit. Two labourers got buried under this heap of loose earth, and before help could be arranged for their rescue, both died due to lack of oxygen.

Learning: We must ensure that the said contractor adheres strictly to all the conditions stipulated in the work permit. Had the contractor arranged to remove the loose earth alongside the digging and erected the barricading, this fatal accident could have been avoided easily. It is also essential to appoint a knowledgeable supervisor on-site to take care of all safety regulations.

3.19 Handling of oil at auto-ignition temperature

Auto-ignition temperature of any substance is the spontaneous (self-ignition) temperature at which the substance catches fire without any external ignition source. Generally, heavier petroleum products have lower auto-ignition temperature. Products with vacuum residue and long residue can self-ignite at a temperature of around 200 °C, while gasoline (petrol) ignites at a temperature of around 250–280 °C. Hence, refiners and other users of petroleum products have to be extra cautious while dealing with these products. Any leak in the system, where product temperature is more than its auto-ignition temperature, will immediately catch fire.

3.20   Pyrophoric iron on fire

Pyrophoric iron is a self-igniting substance, which ignites the moment it comes in contact with air. Normally, it is formed in oil tanks. The rust, formed by corrosion of the inner surface of the oil tanks, reacts with sulphur in the crude oil, in the form of hydrogen sulphide gas, to then form iron sulphide, which is pyrophoric. This compound when exposed to atmosphere on drying self-ignites and catches fire.

I remember, one day, I got a call from the oil movement and storage area of a refinery, informing me that the sludge removed from an imported crude oil tank (containing sulphurous crudes) had caught fire. This refinery was using indigenous crude oils since its inception. These were sweet crude oils and no such incident had taken place thus far. It had, however, started processing imported crudes after the installation of a new distillation unit a few years ago. The crude oil tank containing these imported sulphurous crudes was being cleaned for the first time. The drying sludge began to burn, and caused and explosion.. The staff had forgotten that this product could cause explosion during the unloading of a crude oil tanker, if exposed to air.

Learnings:

* To protect it from ignition, always keep it wet or surround it with a nitrogen atmosphere.
* The sludge removed from tank should be kept under a layer of water and buried at a safe place, or burned under controlled conditions.

3.21   Oil-soaked insulation on fire

To transfer heavy and viscous products through pipelines, these are pumped in heated condition, and to maintain them in fluid state, insulation is done around the pipeline. In some products, steam tracers are also laid along with the pipeline, and over the tracers, insulation is done to keep the product heated. When there are a number of pipelines running together on pipe-rack, sometimes due to a minor leak, the insulation of one pipeline may get soaked with the product. Depending on the auto-ignition temperature of the leaking product and temperature of the line on which it has been falling, the oil-soaked insulation catches fire. The only way to extinguish this fire is to remove the effected oil-soaked insulation completely and put a lot of water on it to cool it down. Many time, the insulation may catch fire again as the oil present in the inner layers seeps out slowly and catches fire again. So, we have to be very careful while tackling the insulation fire.

3.22   Accidents during turn around of processing units

Carrying out maintenance work in a refinery is often more unsafe than the project site. Even though the columns, pipelines and connected vessels are made hydrocarbon free by draining and steaming, some oil or gas remains trapped in some remote corner of the vessels. Giving any entry permit for inspection or hot work in vessels is very risky under such circumstances. One must ensure by meter-check of explosives and gas meter tests that the vessels are absolutely hydrocarbon free. For the sake of safety, one person should always be present outside the vessel to provide immediate assistance to the person inside. I have come across casualties by asphyxiation inside the vessel, or due to explosions triggered by the presence of some oil/gas pocket inside. This cautionary measure is especially necessary for products like LPG, gas, naphtha, motor spirit and kerosene, etc.

In a unit shut-down, many agencies work in a small area and are under a lot of pressure to complete the shut-down in a minimum number of days to minimise production loss. This increases the risk factor.

In one of the refineries, I applied some innovative methods in consultation with the maintenance team:

1. I had seen small external cranes being used in civil construction sites to carry both material and manpower to higher levels. We installed a similar lift against the main column platform with intermediate connections at 3–4 levels. This saved us a lot of time in moving the scrap from higher levels and in the lifting of fresh material to the desired levels. The lift was also used to transport manpower up and down so that they were not tired out..

2. Normally, after water washing and steaming, the columns/vessels are hot and stuffy. We installed cold air blowers in the bottom manhole of the main column to provide fresh and cool air to manpower working inside.

3. To prevent manpower leaving the columns, or coming down from working platforms, arrangement was made to provide tea, snacks and water at the work-spots. These measures saved us a lot of productive time.

4. The lift facilitated the safety, inspection and maintenance manpower to frequently make visits and supervise the job closely.

5. To work on pipelines, instead of erecting scaffoldings at a number of locations, we developed a moving platform on wheels, which

again saved us time and effort of erecting a number of scaffoldings at different locations.

6. Proper communication system was provided to establish an easy communication channel between people working at elevation and ground level.

I am happy to say here that in this total refinery turnaround of about 50 days, there was no lost-time accident.

3.23   Chlorine gas cylinder leak

Chlorine is used in many industries as bleaching agent, solvent and in water treatment plant, etc. Chlorine was used in the paper mill where I worked for few months. It comes in cylinders weighing about 650 kg, which is stored in a segregated area. It is yellowish in colour, heavier than air and has very pungent smell and suffocating odour. The prolonged symptoms intolerable to patient might lead to possibilities of pulmonary embolism, and in case of acute exposure, it may lead to acute lung injury and in some cases leading to death.

Chlorine was being used in the paper mill to bleach the pulp so as to give paper its white appearance and to remove 'lignin' an element of wood fibre that yellows paper when exposed to sunlight.

One day, a chlorine gas leak emanated from a cylinder and started increasing rapidly. There was panic in the mill as gas, being heavier than air, started spreading in other parts of the paper mill. No one had the courage to go near the leaking point and plug it.. Management advised people to stay away and use a handkerchief to cover their noses and advised the fire-fighting team to spray lots of water on the cylinder and gas to prevent it from spreading everywhere. Spreading of foam on the leaked compressed gas further reduced the evaporation. The cylinder was then lifted with the help of a crane and put in the neutralisation pit containing solution of caustic soda. This brought the situation under control and normal activities could resume after a struggle of six hours.

Do's and Don'ts for handling gas leak emergency

- People should stand in upwind direction
- Ventilate the closed space
- Immediately call emergency numbers of the region
- Isolate leak area for at least 100–200 m in all directions by putting road blocks

- Don't panic and keep a handkerchief on the mouth; keep breathing normally
- At least two persons should go to attend/arrest leak. The use of Self-contained Breathing Apparatus (SCBA) is essential prior to going towards the leaking point
- If all efforts to control the leak fail and leakage continues, neutralise the chlorine by passing it into a solution of caustic soda or soda ash or hydrated lime through a suitable pipeline with a perforated distributor
- Standard protective equipment, emergency equipment, standard first aid and emergency health management procedure should be attempted
- The industry must have a chemical disaster management system to tackle a major leak, which may even affect the neighbourhood community.
- A well prepared standard kit to attend chlorine cylinder body leak and knob leak should be kept ready at all times.

3.24   Improper stacking of pipes

The refinery industry needs large-diameter pipes for transportation of raw water, cooling water, effluent and at some places even for the feed and products. Stacking of these pipes in an orderly way is very important from the safety point of view.

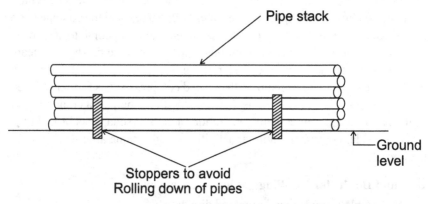

In one refinery project area, the 12-inch-diameter steel pipes were stacked. One crane was lifting pipes from the stack and loading them into a truck parked nearby. The technicians were putting the sling on the pipe and co-ordinating with the driver to lift it. One contractual supervisor was standing near the stack. Suddenly, the stack got disturbed and the dislodged pipes started rolling down. The supervisor tried to run and save himself but still some pipes hit him in his foot, which was badly hurt. Had he delayed even a bit, he might have been crushed under these pipes.

Learnings:

- In stacking of pipes, we must use some stoppers at each stage to avoid any rolling down
- Chaining of pipes after two rows of stacking will also help in preventing the rolling
- At the bottom of the stack, we should put some stoppers, in case rolling starts (the stopper is a solid pipe of around four-inch diameter, which stands two feet in the ground and three feet above the ground and located around two metres away from the stack)
- All the persons working in that area should be on the side of stack or beyond the stoppers
- Wearing of safety shoes is a must in this area

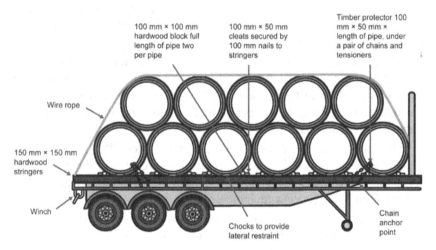

This is the correct procedure for loading of pipes on trolly

3.25   Explosion and fire in gas processing plant

Brief description:

Recently, an explosion followed by major fire occurred near Oily Water Sewer (OWS) in a gas processing plant at Maharashtra, leading to the death of four persons. Heavy rain in area led to an overflow of OWS pit in the plant and spread of hydrocarbon (HC) vapour. Vapour exploded, a major fire followed, leading to four fatalities and damage to the plant.

On the day of the incident, heavy rain had fallen. Security personnel noticed the distinctive smell of hydrocarbon near the OWS pit area. There was HC vapour in storm water channel, which had overflowed. A fire-fighting crew equipped with SCBA sets were engaged alongside operators to identify the source of the HC leak. Suddenly, there was an explosion in the HC vapour followed by the eruption of a major fire. The unfortunate incident led to four fatalities. The plant was damaged badly. Fire was extinguished within a few hours by the fire-fighting team of plant with mutual-aid partners.

The gas processing plant had been receiving gas from offshore pipelines. Pigging operation (a form of flow assurance where pipeline pigs are used to clean pipelines to keep them running smoothly) in this gas line was carried out a few months back, but no pig was found after opening the barrels at plant end. About a fortnight days prior to the incident, high condensate/muck flow was observed inside the gas pipeline due to sudden pig movement. This unexpected event resulted in a huge amount of condensate/muck routed to process units. In many units, OWS pits were reported to be choked with muck. This muck was collected in drums, and filter elements choked with oil-laced muck were also reported lying in unit area.

The plant had provision of routing storm water from gas processing units to OWS pit, if contaminated with oil. All the associated units storm water drain was lined up to OWS pit at the time of incident. OWS pit Pump-A was out of order. Pump-B was put on load but no reduction in pit level was observed. Trend of OWS pit showed pit overflow.

Detectors were detecting HC and giving alarm intermittently in one of the gas process units on the day of incident. Probable causes of the presence of HC in this unit could be the release of HC from the muck due to heavy rain or leakage from pump/process equipment.

At the time of incident, there was condensate in storm water channel due to overflow of OWS pit. The team of fire-fighting as well as operation was trying to identify the source of leakage.

At around 07.00 hours, there was explosion followed by major fire. The fire extended outside plant area as well as to around 1.2 km open channel to sea. Fire was extinguished within few hours.

Fire damage was observed in security cabin, parked vehicles at incident site, around OWS pit area, overhead cable trays, storm water channel of connected unit, etc.

Cause of explosion/Fire:

Root cause of explosion and fire was primarily due to the loss of primary containment caused by:

- Failure of both the OWS pit pumps during monsoon period, leading to overflow of light condensate from pit.
- Presence of HC muck in the unit because of which rain water was diverted to OWS pit. The fire got further spread due to the presence of HC-laced muck in storm water channel.
- Suspected leakage from one of the units where HC detectors had activated intermittently prior to incident.

Conclusion:

OWS pit receives condensate from process units and pumps it out to Effluent Treatment Plant (ETP). In OWS pit, Pump-A was out of order and no reduction in level was observed when Pump-B was put on load. Storm water channel of the respective unit was also lined up to the OWS pit. Heavy rain and non-availability of pumps resulted in an overflow of the OWS pit. Condensate in storm water channel was also a result of OWS pit overflow. As observed from CCTV footage, the fire started with the engine of the vehicle parked at the incident site starting or while fire-fighting crew entered their vehicle.

Recommendations:

- Pit pumps of OWS pits should be maintained in healthy condition.
- In case of HC leak, no vehicular movement should be permitted and affected area should be condoned off.
- Adequacy study of storm water system should be carried out.
- HC muck should not be stored in the plant area and should be disposed off as quickly as possible.
- In the absence of OWS pit pumps, an alternate arrangement for HC removal from the pit should be provided.

3.26 Fatal accident due to harmful gas in an eastern sector refinery, India

Brief description:

A fatal accident of a field operator occurred in the Sour Water Stripper Unit (SWSU) of a refinery due to the inhalation of hydrogen sulphide gas.

At the end of a night shift, in early morning hours, the field operator went inside his unit to collect the samples of sour water, stripped water, rich amine and lean amine from the sampling point. He did not come back with samples. The morning shift field operator went in to check on the operator.. He rushed back to the control room as his personal H_2S monitor started showing the presence of H_2S gas (20 PPM) near the sample collection location and he did not have respiratory protection equipment. After putting it on and returning to the location, he found the night shift operator lying in unconscious condition near the sampling location. He informed the shift-in-charge. Subsequently, he pulled the unconscious operator away from the site. The victim was shifted to hospital, where the doctor declared him brought dead.

In SWSU, sour water from FCC, CDU and DCU are received in degassing drum. The drum floats with acid gas flare system. Depending upon the level of hydrocarbon separated in the reservoir of degassing drum, as per extant practice of the refinery, oil is drained time to time in OWS through a funnel. Sour water is fed to sour water stripper column for the removal of H_2S. The stripped water is consumed in de-salter of CDU. The excess quantity, if any, is drained to OWS.

There were occasions when stripping was found inadequate, which is corroborated by laboratory results that indicate the presence of H_2S beyond the stipulated limits. Hydrocarbon from degasser drum was also drained into OWS funnel instead of CBD. However, the liquid drained overflowed from the OWS funnel, indicating choking in the line.

Other findings at the site include:

- H_2S level in vicinity was in the range of 40–50 PPM.
- The reading of the level transmitter draining water on the degassing drum was found at zero (–2.9). Under such circumstances, if the valve is left open, H_2S inevitably would escape.
- The H_2S detector, near the HC drain funnel, was found non-functional.
- The operator who had gone for sampling did not carry H_2S personal meter.

Cause of incident:

The abnormalities and root causes were as follows:

- Occasional increase of H_2S level (>10PPM) in the stripped water due to improper operation of stripped sour water system.
- Draining of water from degasser drum or any H_2S-contaminated water to OWS manhole, fitted loosely with metal box-type cover.

- It was possible that H2S might have escaped to OWS system from the degasser drum since the level of drum was zero; the drum floats with acid flare gas.

- Lot of H_2S gas escaped through OWS system, since the drain line was choked.

- Operator did not carry personal H_2S monitor.

- The H_2S detector alarm in DCS was either overlooked or kept at the bypass mode.

Learnings:

- The feed to sour water stripper is based on LIC control and must be changed to FRC control system to ensure a steady flow of water to stripper.

- No H_2S-contaminated liquid should be drained into the open system; instead, it should be drained to CBD. On these lines, any access-stripped sour water should be drained to CBD only.

- Entry in process unit with H_2S personal meter should be strictly implemented.

- Additional H_2S detector should to be provided near OWS manhole.

- Whenever any operator goes to the unit, he should inform his co-operator in advance.

- Safety interlock should be made functional and kept on the line.

- H_2S alarm must be acknowledged by control room, and appropriate action taken.

3.27 Fire and explosion in flixborough (UK)

The case studies I have discussed so far are mostly from Indian industries. I have taken up a few in order to point out the general mistakes that cause accidents/fires and explosions.

I have gone through some case studies and videos of accidents that took place abroad. The Flixborough incident was also similar to the other incidents described thus far. There were six reactors through which cyclohexanone had to pass through at high temperature and pressure in order to get converted to caprolactum. Cyclohexanone is highly inflammable; any leak was dangerous and prone to catch fire.

Out of the six reactors, one reactor was put on by-pass for repair work and a 20-inch-diameter by-pass line (jumper) was provided to keep the unit running.

There were frequent upsets in the unit due to several leaks in a short span of two months. Due to several thermal expansions and stresses, the by-pass line failed, and there was a massive explosion followed by huge fire.

The main cause found was the design mistake in putting the 20-inch by-pass line between the two reactors, which was not adequately supported, and the frequent leaks in units caused the fatigue and failure of this piece.

Learnings:

- Management should recognise the vulnerability in critical manpower changes.
- Any modification in plant should be designed, constructed, tested and maintained by the same standards as the original plant.
- All pressure systems containing hazardous material should be inspected by the inspection department after any significant modification and before the system start-up
- The system should be tested at about 1.3 times the relief valve pressure setting.
- The by-pass was tested pneumatically. In such systems hydraulic testing should be made obligatory.
- HAZOP is required to be carried out at regular intervals or whenever any modification is carried out to understand the risk involved.
- Proper supports should be provided even if the line is put temporarily to avoid vibrations/stress and ultimate failure.
- There should be a proper policy for 'management of change' 'to take all precautions and carry out checks at each stage'.

A similar case occurred in another refinery where explosion and fire took place due to a failed outlet pipe of flare knock-out drum, releasing massive hydrocarbons into the atmosphere. This formed an unconfined vapour cloud, which found a source of ignition and exploded. It seems that due to malfunctioning of control valve for pumping out the bottom product from column A to column B, the column A got filled with liquid. This liquid found its way to flare the knock-out drum, and due to an increase in pressure and flow, the weak point in flare header failed and the hydrocarbon vapour escaped.

In this case, it is very clear that the operator could not understand the magnitude of the problem looming in the plant..r. In fact, it was safer for him and the plant unit to shut down and analyse the problem and restart the unit after resolving it. The learning from this case study would be:

- The operator should know how to carry out simple volumetric and mass balance check.
- Staff should be trained based on assessment of their knowledge and competence on handling operations and emergencies.
- There should be a proper inspection plan to check the health of pipelines and vessels.
- Time to time checking of levels and pressure, etc are to be followed by operators in the field.

3.28 Explosion and fire in olefins production unit of formosa plastics corporation in point comfort, texas

Brief description:

On 6.10.2005 at about 3.05 PM, there was an incident of release of propylene from olefins unit-II of Formosa Plastics Corporation, Texas, as a trailer being towed by Forklift snagged and pulled a small drain valve out of the strainer in a liquid propylene system. Escaping propylene vapourised, forming a large vapour cloud. At about 3.07 PM, the vapour ignited creating an explosion. Explosion knocked down several vessels and burnt two (one seriously) operators present in the unit. Flames from fire reached more than 500 feet higher in air. The fire in unit continued for five days. Because of massive fire, evacuation started from sites. Fourteen workers sustained burn injuries. Due to extensive damage, unit remained under shut-down for five months.

The olefins unit-II uses furnaces to convert either naphtha, or the natural gas derived feedstock, into a hydrocarbon mixture containing methane, ethane, propane, propylene and various higher hydrocarbons. Distillation columns then separate the hydrocarbon mixture. Some of the separated gases liquified and sent to storage, while others are used as fuel for the furnace or recycled into feed stock.

Relief valves protects various process equipment from overpressure. These valves discharge into the flare header system, where the hydrocarbons can be safely burned.

Incident sequence:

- A worker driving a fork truck towing a trailer under a pipe rack backed into an opening between two columns to turn around.

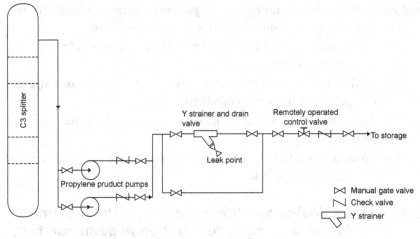

Propylene product flow

- When the worker drove forward, the trailer caught on a valve protruding from a strainer in the propylene piping system.
- The trailer pulled the valve and associated pipe out of the strainer, leaving 1.9-inch-diameter opening.
- Pressurised liquid propylene escapes through the opening and partially vapourised creating both a pool of propylene and a rapidly expanding vapour cloud.
- The fork truck driver and other contractors saw the leak and evacuated the site.
- An operator also heard and saw the release and informed the control room. The control room operator also saw it on CCTV and began to shut down the unit.
- The field operator attempted unsuccessfully to reach and close manual valve to stop the leak. They operated the fixed water monitors.
- Control room operators shut off pumps from MCC and closed control valves to slow the leak.
- The vapour cloud ignited.
- Field operator left the unit.
- Control room operators declared a site-wide evacuation. On getting propylene smell, they evacuated the control room.
- A large pool fire burned under the pipe rack and sides of an elevated structure that supports various process equipment, such as vessels, heat exchangers and relief valves.

- Company Emergency Response Team (ERT) arrived and took command of incident response.
- After thirty minutes into the event, the side of the elevated structure collapsed, crimping emergency vent lines to the flare header.
- Crimping pipes and steel, softened from fire exposure, led to multiple ruptures of piping and equipment and loss of integrity of the flare header.
- ERT isolated fuel supply sources where possible, and allowed small fires to burn the uncontained hydrocarbon.
- It took five days for the fire to be extinguished.

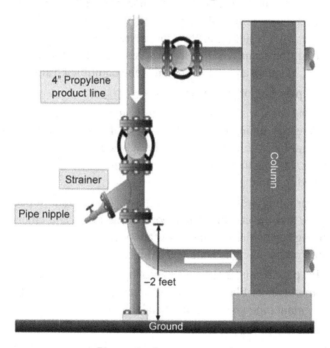

Pipe and valve arrangement

Lapses and cause of incident:

1. **Vehicle impact protection**

   The propylene piping involved in this incident was protruding outside into the open, yet it was not given any impact protection.

   Company had administrative safeguards for vehicle operation in the unit, including a plant-wide speed limit, a vehicle-permitting process, and a crane-use procedure. However, these safeguards did not specify where vehicles were to operate within the unit.

The plant design drawings designate specific access paths for vehicles; however, these were not physically demarcated in the unit itself. The area where the impact occurred was not a designated access pathway, but was large enough for vehicles to pass through.

The guidance manual dealing with the protection of control stations, pipelines, and other grade-level plant equipment was not specific, but it did state that protective measures should be in place to prevent impact.

During facility siting analysis, the hazard analysis team discussed this aspect. They judged the consequence as 'severe' and probability as 'very low', resulting in 'low overall risk range'. Because of lowrisk rating, the team considered that existing administrative safeguards were adequate and did not recommend additional traffic protection.

2. **Fire proofing of structural steel**

During the fire, part of the structure supporting the relief valves and emergency piping to the flare header collapsed. The collapse caused several pipes to crimp, it was likely this prevented flow through the pipes, leading to the rupture of major equipment and piping that only added fuel to the fire.

Fireproofing was done on only three, out of four support column rows, and the column that supported the PSVs and emergency vent piping had no fireproofing. The bare steel column bent over, while the fireproofed columns remained straight.

API publication 2218, 'Fireproofing practices in Petroleum and Petrochemical Plant' (July 1988) recommends that steel supporting important piping such as relief and flare lines be fireproofed.

3. **Remote equipment isolation**

The leak occurred between manual control valves and remotely operated control valve. While a check valve and remotely isolation valve downstream of the leak prevented the backflow of propylene from the product stage, operators were unable to reach the manual valves capable of stopping the flow from the distillation column. The operators were also unable to reach the local control station to turn off the pumps supplying propylene, although they eventually turned off the pumps at the MCC located in the control room building, slowing the rate of propylene feeding the fire.

Had a remotely actuated valve been installed upstream of the pumps, this incident would likely to have ended quickly.

This was in line with designer (M/s Kellogg) philosophy. However, Health and Safety Executives (HSE) (1999) recommend that large vessels and columns with hazardous inventories be equipped with rapid isolation capability.

4. **Flame-resistant clothing**

Flame-resistant clothing will limit the severity of burn injuries to workers in plants where flash fires may result from uncontained flammable liquids and gases. Neither of the two operators burned in this incident was wearing FRC: had they been, their injuries would likely have been less severe.

The OSHA as well as NFPA has guidelines of the use of FRC.

Learnings:

- **Hazard review:**

  When performing a Hazard analysis, facility siting analysis, or pre-start-up safety review, vehicle impact and remote isolation of catastrophic releases should be investigated. Such critical projected equipment should have a permanent impact guard.

- **Flame-resistant clothing**

  In process plants with large flammable liquids and/or gas inventories, mechanical failures can result in flash fires that endanger workers. The use of FRC may limit the severity of injury to employees who work in plants with large inventories of flammable gases and liquids.

- **Use of current standards**

  Evaluate the applicability and use of current consensus safety standards when designing and constructing a chemical or petrochemical process plant. This should include reviewing and updating earlier designs used for new facilities.

3.29   Fatal accident in a petroleum refinery during hydro-jetting operation

Brief description:

A fatal accident of a contractor workman took place on 1.10.2013 during hydro-jetting operation in a refinery for removal of debris from OWS manhole/

lines and ensuring thoroughness. During the job, the self-propelling nozzle of machine came out abruptly from OWS manhole. The water jet and nozzle hit the neck of the worker standing by side of manhole. He was badly injured. Later, he succumbed to his injuries.

What led to this fatal accident? Let us analyse the incident. The truck mounted hydro-jetting machine with four workmen (including a supervisor) reported at site for de-choking of OSW manhole (Depth: about 21 feet) and ensure thoroughness between two manholes. Job was started after tool box talk and obtaining work permit (cold work) and continued till 1.30 PM, till a leak developed in discharge line of the hydro-jetting pump. The job discontinued, and the truck was sent out for repair by contractor.

Truck-mounted hydro-jetting machine had a pump driven by truck engine through a Power Take-Off (PTO) unit for hydro-jetting. The pressure for hydro-jetting was controlled through a lever on rear side of the vehicle. The discharge hose of pump was connected to a self-propelling nozzle.

 High-Thrust Conical Head Nozzle	High-trust conical head (self-propelled) nozzle used for the purpose had one hole in front and five holes at the rear. Hole in front is used for de-choking/cutting/cleaning, whereas holes in the rear provide the thrust so that nozzle moves in forward direction automatically.

After repairing the machine, the contractor started the job without informing the engineer in-charge, while taking clearance in work permit on the same day (afternoon). Two contractor persons were at the work site. When job was started, the manhole had about three feet water and the OWS pipes were not visible from the manhole top. During the job, one person was at pump control (on rear side of vehicle), whereas another was trying to put the nozzle in OWS pipeline (covered under water) by swinging the hose under pressure (300–500 PSI) from top of manhole. Such nozzle with hose has tendency to turn back, if it finds an obstruction in the middle of the line, provided there is sufficient space for turning back. In this case too, nozzle under water might have been obstructed under the water and turned back as there was enough space for doing so. Thus, after turning back, the hose, in a swift motion, moved in the direction of the operator. Consequently, nozzle with hose at a pressure of about 300–500 PSI hit the neck of the operator, causing fatal injuries.

Lapses/cause of incident:

1. As the contractor personnel entered premises with a temporary gate pass of a single day, they skipped the structured safety training. The tool box talk given at site did not address the relevant hazards and precautions.

2. Job was being done without supervision (from company side as well as contractor side).

3. Job was re-started without taking clearance on work permit for the second shift.

4. SOP and safety precautions for high-pressure jetting system issued by OEM were inadequate and not available at site.

5. SOP provided by the contractor for the job was not reviewed for the content.

6. PPEs were not categorically identified for hydro-jetting.

7. Insertion of the hydro-jetting nozzle inside OWS pipe was being done by swinging the hydro-jetting hose, which s is an unsafe method.

   'Why Why Analysis' of Incident concluded that ignorance about the hazard/risk associated with the job by all concerned leads to the fatal accident.

Learnings/recommendations:

• System needs to be developed for assessing the hazards/risks associated for hydro-jetting of various process equipment/facilities.

• Compliance of work permit is to be ensured in line with OISD Std.-105.

• For high-pressure system of contractors, the test certificate should be obtained from contractor and all repair and maintenance work of system should be subjected to test, recommended by OEM, before taking equipment into service.

• In-house competency needs to be developed to supervise the execution of such jobs safely. Contractor must depute a dedicated supervisor for such critical jobs.

- Existing access control system to be reviewed. All entry/exits of visitors/employees/contractor personnel should be linked through the access control system.
- Structured safety induction training should be made mandatory for all contractors, coming for the execution of job inside refinery, even if for one day or two.
- Training programme needs to be conducted for creating awareness amongst permit signatories for critical operations such as using high-pressure systems.
- For such specialised jobs, requirement of PPE should also be identified and arranged, and compliance should be ensured.

3.30 Fatal accident at a construction site in a petroleum refinery due to fall of scaffold pipe from height

Brief description:

On 11.11.2017, during the revamp job of the DHDS unit in a petroleum refinery, a fatal accident of a contractor employee took place, when he was hit on the head by a falling scaffold pipe from a great height.

During revamp of DHDS unit, two contractors were allowed to carry out work at different elevations in a reactor at a height of 5M at the lowest platform and 27M at the top platform. The erection of reactor was already completed. On 11.11.2017, one of the contractors, who was working at 27M height, was carrying out structural work for platform/bracing on the southern side. However, another contractual employee was carrying out hot job for the erection of cable tray for laying instrument cable at 5M level on the northern side. Two contractor supervisors were supervising the erection job of cable tray at 5M level from ground floor, standing below the job.

At the top platform of reactor, a large portion of the scaffold was jutting out on an extended/cantilever platform on another side. Three vertical pipes of scaffold were hanging outside the top structure, supported with a scaffold coupler. One of them loosened, slipped and fell from a height of about 27M. The falling pipe struck the safety helmet of one of the contractor supervisors, who was standing directly below, having breached the soft barricade on the ground floor. The impact of scaffold pipe damaged the safety helmet of supervisor and caused a critical head injury. He was shifted to hospital immediately, where he succumbed to injuries.

Lapses and cause of accident:

1. Original scaffold for work at 27M level of reactor was erected from ground level. Later, some portions of scaffold were removed, leaving part of the scaffold hanging from the respective beam of extended platform/cantilever platform. These scaffolds were left hanging to carry out localised jobs later. In horizontal pipes, beam clamps were used, whereas for vertical pipes, scaffold couplers with bolts were used.

2. A scaffold pipe of 3M (weight about 12 kg) loosened and fell down. Why?

   The scaffold pipes were kept hanging at this height since about two months, and due to self-weight, wind load and temperature effects, the bolt on clamp might have loosed/threads worn off, leading to slippage of pipe from the location and falling down.

   Neither the contractor nor the Project Management Consultant (PMC) had any record of scaffold pipe and coupler checking/testing. Record of certification and re-certification of scaffold were also not available. *In brief, the scaffold safety standard was not followed for erecting, dismantling, testing and certification of scaffolds.*

3. The contractor as well as PMC failed to follow and implement the precautions mentioned in work permit and job safety analysis (JSA) at work site. This also indicates poor supervision of job.

4. The supervisors of contractor for working at 5M level were standing within the barricaded area, ignoring the safety warning.

Learnings/Recommendations:

- Scaffold safety standard is to be implemented at site by contractor as well as PMC in totality. Other safety precautions are also to be implemented.
- Scaffold certification/re-certification should be approved by either PMC or independent scaffold inspector.
- Separate specialised agency/group is to be engaged for ensuring compliance of scaffold safety norms, training of site personnel and regular check of the fitness of scaffold.

3.31 Explosion in BP refinery, texas

An unconfined vapour cloud explosion took place in British Petroleum Refinery, Texas, on 23rd March 2005 during start-up activity of Isomerisation

(ISOM) process unit, followed by massive fire and explosions, killing fifteen workers, injuring about 170 others and severely damaging refinery. The detail investigation report by Chemical Safety Board (CSB), USA, is available on website www.csb.gov>bp-america-refinery-explosion.

Brief description of incident:

ISOM unit in refinery was meant for conversion of low-octane hydrocarbon through various chemical processes into hydrocarbon with higher octane rating that could then be blended into unleaded gasoline. The raffinate splitter of the unit had a 170-feet-tall splitter column, used to separate lighter hydrocarbon components from top of the column (mainly pentane and hexane), which condensed and were then pumped to the lighter raffinate storage tank, while the heavier components were recovered from lower down in splitter and then pumped to heavy raffinate storage tank.

Two turnaround activities were taking place at the adjacent Ultra-cracker Unit (UCU) and at the Aromatics Recovery Unit (ARU) at the same time.

After work had been completed on the raffinate splitter, the start-up activities began. One of the primary safety critical steps in the pre-start-up process was the pre-start-up safety review procedure. This was not carried out probably because the unit had many serious safety issues.

Start-up activity commenced in the night of 22nd March with filling up of raffinate column of unit. Later this was discontinued but re-commenced at 9.30 AM on 23rd March. The level transmitter of column was designed to indicate the raffinate level within 5-feet span from the bottom of the splitter column to a 9-feet level. Before re-commencing the column fill and circulation process, heavy raffinate was drained from the bottom of the column via the LCV into the heavy storage tank and was then shut off in a manual mode with a 50% flow rate.

The level in column started build up as there was no circulation and feed to column continued. The defective-level transmitter of the column continued to show the level at less than 100% (at 9 feet) and external sight glass was opaque.

Further burners in the furnace heating the feed raffinate led to high temperature and pressure in the column. The level in column rose had to 98 feet. At about 12.42 PM, the furnaces had been turned down and the LCV was opened. The level in the column went up to 158 feet. At about 1.13 PM, all the three PSVs were forcibly opened due to high pressure, and led to heavy flow of heated raffinate. Hot raffinate flowed in to blow-down drum and stack, and as it filled, some of the fluid started to flow into the unit sewer system via a 6" pipeline at base of blow-down drum. HLA of drum also did not activate. After sometime, the hot raffinate started falling from top of the stack and into the air. The hot raffinate rained and spread, making pool of oil.

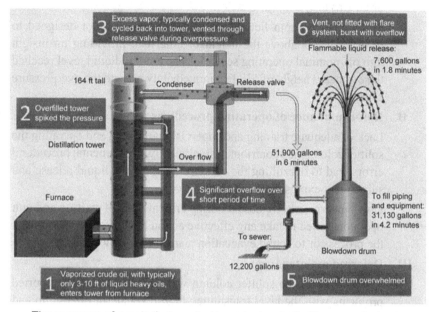

The sequence of events that resulted in explosion, ignited by a near by truck.

The vapour cloud of hot raffinate spread and reached a diesel pickup truck trailer, located at about 8M away from the blow-down drum in start position. Getting source of ignition, the vapour cloud exploded. The blast pressure wave struck the nearby contractor shed, destroying it completely, killing fifteen people and injuring about 170. The explosion was followed by massive fire and further explosion. Incident led to severe damage to ISOM unit and unit remained out of operation for the next two years.

Cause and contributory factors leading to explosion:

Personnel responsible for start-up greatly overfilled the splitter column and overheated its content, which resulted in an over-pressurisation condition. Liquid was pumped into the column for almost three hours without any liquid being removed or any action taken to achieve the lower liquid level mandated by the start-up procedure.

The contributory factors were as follows:

I. **Inadequate design**

BP decision to continue operating with an atmospherically vented blow-down stack in lieu of the widely available, and inherently safer, flare stack was a contributory factor. The capacity of the blow-down drum was also a limiting factor.

The splitter column liquid-level transmitters were not designed to measure levels above the height of nine feet, providing no insight into off-nominal operating scenario. The column liquid level reached the estimated height of 138 feet immediately prior to the overpressure event.

II. **Non-compliance of operating procedure:**

Lack of adequate training and supervision of filling and operating the splitter column had contributed significantly. Fundamental procedural errors lead to overfilling the column, overheating, liquid release and subsequent explosion.

Unit supervisors were absent during critical parts of start-up and unit operators failed to take any effective action to control deviation from the process or to sound evacuation alarms after PSV opened.

III. **Deferred maintenance:**

The start-up of the splitter column was authorised despite reported problems with the level transmitter, high-level alarm of column and blow-down drum. These had management approval. A key alarm failure within the column and blow-down drum failed to warn the operators.

IV. **Trailer in hazardous zone:**

Mostly, fatalities occurred in and around the trailer. Trailer was parked in hazardous zone near to the stack and blow-down drum. This was violation of facility siting policy.

Further, pre-start-up safety review was not carried out, and hence, the trailer and contractor sheds remained in operating area before start-up activities.

Learnings/Recommendations:

- Ensure pre-start-up safety review before each start-up.
- Ensure that instrumentation and process equipment necessary for safe operation is properly maintained and tested.

- Effective implementation and control system for splitter column such as multilevel indicators and automatic control. Configure control boards display to clearly indicate material balance for stripper column.
- Carry out details PSM (including MOC and HAZOP) and ensure their implementations of recommendations.
- Introduce system of Process Safety Performance Indicator (PSPI) for monitoring of each leading and lagging process indicator.
- Ensure training and skill development of operators using simulators.

3.32   Explosion inside flammable solvent storage tank due to static electricity

On 17.07.2007, a series of explosions followed by fire occurred in Solvent Tank Farm at Borton Solvents Wichita facility in Valley Center, Karnas, USA. The incident destroyed the entire tank farm and led to eleven people being injured. The area outside the premises was also evacuated. For details, refer CSB website: www.csb.gov

Brief description of incident:

The tank farm handling inflammable solvent had 43 above-ground storage tanks, ranging from 300 to 20,000 gallons capacity. In tank farm, the Varnish Makers' and Painters (VM&P) naphtha, a flammable solvent, was being received through tank trucks and being unloaded through pumps in storage tanks.

The salient properties of VM&P naphtha, which indicate fire hazard, are as follows:

Flash Point	14 °C	
Vapour Pressure	0.7 kPa (5 mmHg) at 20 °C	
Flammability Limit	0.9–6.7% by volume in air	
Conductivity to Static charge	3 pS/m	Non-conductive

On the date of incident, the unloading of an above-ground storage tank holding 15,000 gallons of VM&P naphtha was in progress via a pump from a tank truck. The tank truck had three compartments. The unloading from the first-two compartments was completed and the unloading from the third compartment had just begun. At this point, the first explosion occurred in receiving naphtha, sending the tank rocketing into the air, trailing a cloud of smoke and fire from burning liquid. It landed approximately 130 feet away. Within moments, two more tanks ruptured and released their contents in tank farm. As the fire burned, the contents of other tanks over-pressurised or ignited, damaging tanks one by one.

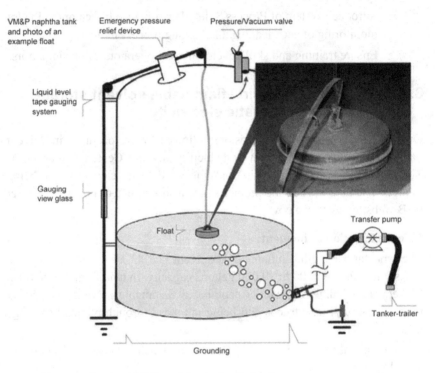

Diagram showing VM&P naphtha tank with floating gauge arrangement

Cause of explosion and contributory factors:

- The tank contained an ignitable vapour-air mixture in its tank head space.
- Stop-start filling, air in the transfer piping, and sediment and water (likely to have been present in the tank) caused a rapid static charge accumulation inside the VM&P naphtha tank.
- The pump flow velocity was very high (4.6 M/second) for non-conductive liquid. This helped in static charge generation.

- The tank had a liquid-level gauging system float with a loose linkage that most likely separated and generated a spark during filling.
- The turbulence and bubbling during stop-start transfer pumping, in addition to creating rapid static charge accumulation, also in all likelihood created slack to the float, causing the linkage to separate and spark.
- The MSDS for VM&P naphtha involved in the incident did not adequately communicate the explosive hazard.

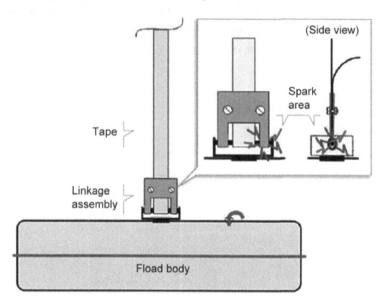

Diagram showing float linkage and area where the spark likely occurred

Lesson learned:

- Add an inert gas such as nitrogen to the tank head space with consent of manufacturer of VM&P naphtha, which will reduce the potential for ignitable incident (explosion) as it renders tank head spaces incapable of supporting ignition from static spark.
- Modify or replace loose linkage tank-level floats, so that floats with level measuring devices do not prompt spark inside the tank.
- Remove any slack in the tie connected to the float mechanism that could allow a spark gap to form.

- Non-flammable liquids capable of forming ignitable vapour-air mixture inside tanks should be transferred at reduced flow (pumping) velocity (1M/second) to minimise the potential for static ignition.

- Anti-static (conductivity-enhancing) additives increase the conductivity of liquids helping reduce static accumulation. This may be added with consent of manufacturer of VM&P naphtha.

- Covering 'earthing and bonding' against static charges alone for protection is not enough in MSDS of VM&P naphtha by manufacturer. It should provide conductivity testing data and specify the additional precautionary measures that should be observed.

Chapter 4
Tips to maintain a healthy safety system

There are some safety tips for all working in industries which are easy to remember and very effective in improving the safety record of any industry. Many such tips can be displayed at various strategic points to catch the attention of staff moving in the field. These are mostly based on experience gained is many years.

Chapter 4
Tips to maintain a healthy safety system

Here are some tips to be remembered to maintain a good and healthy safety system in an organisation:

There is a saying that wise people learn from the mistakes of others. In the context of the subject of safety, past incidents will prove to be our best teachers. Let us learn from past case studies and extract pointers to avoid similar incidents in industry.

- There can never be any relaxation on safety. Wearing PPEs is must for everyone in the field, with no exception to be made for top management and visitors. Safety helmets must be properly secured by tightening the straps over chins or tightening them using the knobs at the back of helmets.

- Reporting of substandard acts/conditions/near-miss incidents is the responsibility of all. Through near-miss analyses, we can know which areas need improvement and take immediate corrective measures accordingly.

- Learning from one's own mistakes is good but learning from others' mistakes is wise. Hence, it is recommended that all case studies be circulated to all concerned, along with learnings from past accidents.

- No visitor is allowed to go inside the plant area without being accompanied by a representative from the host company.

- A drunken person inside the installation is a danger to himself, and others as well as to the plant assets. Alco-meter test should be conducted for all those entering the premises, even if they belong to the top management (my experience of Kenya Petroleum Refinery).

- Static electricity is most dangerous in industries handling inflammable materials/products. Hence, all care must be taken to ensure proper earthing of all vessels, tanks, pipelines and wagons/trucks under loading/unloading. Lack of earthing has been known to be one of the main causes of fires and explosions in the industry.

- The plant manager must do safety rounds of site daily to ensure tidy housekeeping/no obstacle on exits or escape routes He must regularly check the knowledge of operators about the use of PPEs and the handling of every kind of emergency, besides keeping a check on compliances in the case of of work permit, etc.
- Keep a check on open drains of the installation. Any presence of oil in them is very dangerous.
- The plant area needs to be tidy, especially the pump/compressor area. Any object on the way or oil on floor is a definite route to an accident.
- Use of standard safety harness must be ensured to avoid any free fall while at height. No work at height in foggy weather/during rain and storm is to be permitted.
- Suitable safety notice boards displaying do's and don'ts are needed in accident-prone areas, like that of chemical handling and water draining systems from crude oil tanks /product tanks and LPG storage facility. Any negligence there can cause major emergency.
- Many accidents are reported due to failure of scaffolding. Standard scaffolding material should be used. Once it is erected, it should be certified as 'fit to use' by a competent officer.
- Regular health survey of process pipelines will help in controlling the failure/ leakage of pipelines.
- Any leaky point/ dripping point should be attended to immediately.
- Don't allow anyone to walk on pipelines/unsecured overhead platform as any fall may lead to head injury/fatality.
- In case of strong winds/cyclone conditions, the booms of all cranes should be lowered down and brakes to be properly engaged to avoid rolling of cranes (case in point: accident at one of the ports in eastern India , when a heavy crane rolled down into the sea.)
- One has to be very careful (stand to one side) while lighting the cold heater or boiler, as the flame usually backfires.
- While climbing onto the top of a floating roof tank, one should only walk on the platform and not climb onto the roof. Accidents have been reported due to corroded and weakened roof plates falling inside the tank when walked on.
- Safety-conscious people are always alert to danger and keep their nose, ear and eyes always open. Any smell of gas/hydrocarbon

vapours should alert people about leaks and these must be attended to immediately, to avoid any emergency-like situation.

- Heavy oils should not come in contact with hot surfaces like steam lines or steam tracers as there are good chances of catching fire. If these products fall in droplets from overhead lines over a hot surface, they can easily catch fire due to their low auto-ignition temperatures.
- Repair work on electric poles during rainy seasons should be avoided,as chances of electric shock, even electrocution, are great. In cases of emergency where such work is unavoidable, extra precautions must be taken.
- Electrical short circuiting has been one of the common causes for fires in plants. Standard and proper rated electrical fittings/wires/ joints should be used to avoid such accidents.
- Oily sludge taken out during tank cleaning, vessel cleaning, etc. contains pyrophoric iron. This is highly inflammable and can catch fire as soon as it dries . Hence, sludge should always be kept wet or stored under water to prevent exposure to the sun. This sludge should be treated in line with environmental guidelines.
- Dry grass, can being easily ignited, its presence in the plant area is dangerous. It should be cut and removed immediately.
- There should not be any tall trees in the plant area. During storms, they can fall and damage the process pipelines and power cables. They can even obstruct the fire-fighting operation.
- Wet floors or any foreign material lying in the plant area can lead to slip/trip and cause accident. Hence, proper and large caution boards should be placed to alert passers-by.
- All manholes should have proper covers. Any manhole, kept open for some job, should have proper barricading to prevent anyone falling into it. Plant area should be well lighted.
- Training and refresher training of employees is the only way to make them aware about safety and understand the proper use of safety equipment during an emergency.
- Weekly trials of fire water pumps and daily trial of emergency sirens should be practised. Regular mock drills in various types of emergency scenarios will bring out weaknesses in the system, and enable corrective measures to be taken.

- Safety audits by multidisciplinary audit teams consisting of persons drawn from different locations also helps in bringing out deficiencies in the system and to plug them. Most important is to monitor the implementation of the recommendation, by top management in monthly safety reviews.

- Head of HSE department has to be a dynamic leader, and he should report to the location head for effective control.

- Regular and candid reporting of loss time incidents to top management should be encouraged. Any one trying to hide such incidents should be liable for punishment.

- Monthly HSE meeting to be attended by all HODs and top management; is a must to discuss and analyse safety breoaches, their causes and decide on actions to be taken to improve overall safety standard.

- Safety showers and eye wash facilities are must in the chemical handling areas.

- Limited vehicle movement should be allowed in plant area with strict adherence to speed limit. Vehicles fitted with CCE-approved spark arrestor should only be permitted as per requirement of hazardous area classification. Extra care should be taken while material shifting, and hanging objects should have a guided movement.

- System should be in place to involve workers in safety management system through floor-level safety committees, safety meetings, safety training, safety inspection and safety campaigns, etc.

- Safety bulletins should be circulated widely among employees. These should cite case studies, unsafe acts, near-miss incidents, accident-free man hours/days achieved without a loss time accident vs the target. Even some token awards for achieving specific targeted accident-free days may be introduced to encourage the people to follow safe practices.

- Safety performance should be one of the parameters in working out the formula for the annual performance incentives. Also, some incentive schemes can be introduced to honour the employee who does an exemplary job to avoid a major mishap due to his alertness, knowledge and courage. I found it very encouraging when I tried this in one of the refineries.

- All the areas with moving parts should have guards, and no one wearing loose or flowing clothes should be allowed to enter the plant.
- During project stage/annual maintenance, the proper metallurgy of pipes and bends being used should be ensured. Any wrong fitting may result in leakage and thereby, fire during plant operations.
- In project contracts, a clause providing incentives on achieving milestone of accident-free man hours/man days as well as a penalty clause for each accident in project area should be inserted.
- The contractor has to ensure the availability of a qualified safety officer at project work site to oversee safety compliances. It should be the responsibility of the contractor to provide PPEs to its workforce.
- The operating staff needs to be more careful when any interlock control is put on the by-pass. This job should be attended to on top priority, and everyone on duty should be aware of this by-pass. Approval of such by-pass should be done as specified in Management of Change (MOC) procedure.
- The control room should not be left unattended under any condition.
- Predictive/preventive maintenance schedule of each running equipment should be strictly followed to avoid sudden failure of equipment. Similarly, any vibration in pipeline should be given proper support to prevent its fatigue failure.
- Any standing instruction/special attention job should be explained to operating people as soon as they join duty. (The reliever should take full brief from the person in earlier shift, besides reading the instructions.)
- Availability of the first aid box should be ensured at all work sites. Also, an adequate number of 'first aid trained' people should be available during operating hours. It has been experienced that if timely first aid is administered to an injured person, chances of saving his life gets increased considerably.
- Only medically fit persons should be deployed for working at heights.
- Everyone should follow Standard Operating Procedures (SOP)– shortcuts cannot be allowed.
- The staff should not be allowed to work if tired. Long working hours without breaks or proper rest is unsafe. Unless the person is alert on the job, chances of mistakes are high.

- Extra care is required for working in confined spaces/ at heights and during excavation and in any other high-risk job. Proper Job Safety Analysis (JSA) should be done prior to the start of the job to identify probable hazards and their mitigation measures to be followed.
- Work area monitoring with respect to noise and presence of toxic/ hazardous vapours should be practised regularly to avoid any undesirable exposure to working staff.
- Regular health monitoring of working staff should be implemented to assess the occupational health issues.
- All field and portable safety equipment for example detectors, alarms, shut-down switch, and extinguishers, should be maintained by periodic inspection.
- Any major accident can be avoided if the staff on duty is alert, knows the job and trained to handle any emergency. Many times, simply the use of common sense is very helpful in controlling the situation.
  - Once there was fire in a Naphtha wagon under loading, operators decoupled the rest of filled wagons and pushed them to a distance. This averted a major fire.
  - In one unit, sulphur dioxide leaked and caused panic in the refinery. One operator used his presence of mind and entered the gas cloud while wearing air breathing apparatus and closed the valve to arrest the leak. This action prevented a major emergency.

The above tips will certainly be of help in reducing incident, in your industry. I wish all of you and your colleagues a very safe working and long&and healthy life.

Bibliography

1. Frank E Bird Jr., *Domino Theory*
2. Kletz Trevor '*Still Going Wrong*' 2003
3. OSHA Process Management Standard
4. Conoco Phillips Marine Pyramid (Incident)
5. ISO 45001: *Occupational Health & Safety Management Systems*
6. OISD GDN 206: *Guidelines on Safety Management System in petrochemical industry*
7. *Process Safety Management for Petroleum Refineries*, OSHA 3908/3918
8. *Safety & Health Guide for Chemical Industry*, OSHA 3091
9. Investigation report of cases by Chemical Safety Board (CSB), USA
10. Case studies published in safety journals and safety bulletins
11. *Borton Solvents Investigation case*, CSB news release
12. API recommended practices 2003
13. Protection against ignition arising out of static lightening stray currents, 7th edition 2008
14. Presentation '*Learning for process industry from the Bhopal gas tragedy*' Shri S.P. Chaudhary on 11–12 February, 2020 in New Delhi
15. COX J. (2005) Flixborough Revisited Chemical Engineering The Chemical Engineer 26–28
16. The report of the US Refineries Independent Safety Review Panel (Baker Panel) January 16, 2007
17. Fatal Incident Investigation report, Isomerization Unit explosion, find report, Texas City, USA, British Petroleum, Mongford J.
18. Alkylation Unit Ginza Oil Refinery, James Town, NM
19. Report of the investigation by the health and safety executive into the explosion and fires at Texaco refinery in July 1994.

Printed in the United States
by Baker & Taylor Publisher Services